Bioethics and the Fetus

Biomedical Ethics Reviews

Edited by

James M. Humber and Robert F. Almeder

Bomedical Ethics Reviews • 1991

Bioethics and the Fetus

Medical, Moral and Legal Issues

Edited by

JAMES M. HUMBER
and ROBERT F. ALMEDER

Georgia State University, Atlanta, Georgia

Humana Press • Totowa, New Jersey

The Library of Congress has cataloged this serial title as follows:

Biomedical ethics reviews—1983-   Clifton, NJ: Humana Press, c1982-

v.; 25 cm—(Contemporary issues in biomedicine, ethics, and society)
Annual.
Editors: James M. Humber and Robert F. Almeder.
ISSN 0742-1796 = Biomedical ethics reviews.

1. Medical ethics—Periodicals. I. Humber, James M. II. Almeder, Robert F. III. Series.

[DNLM: 1. Ethics, Medical—periodicals. W1 B615 (P)]

R724.B493   174'.2'05—dc19   84-640015
                                        AACR2 MARC-S

Contents

Preface

Bioethics and the Fetus: Medical, Moral, and Legal Issues is the ninth volume in the **Biomedical Ethics Reviews** series of texts designed to review and update the literature on issues of central importance in bioethics today. All of the essays in this volume examine moral and/or legal problems involving human fetal life; summaries of these essays may be found in the text's *Introduction.*

Bioethics is, by its nature, interdisciplinary in character. Recognizing this fact, the authors represented in the present volume have made every effort to minimize the use of technical jargon. At the same time, we believe the purpose of providing a review of the recent literature, as well as of advancing bioethical discussion, is well served by the pieces collected herein. We look forward to the next volume in our series, and very much hope the reader will also.

James M. Humber
Robert F. Almeder

Contributors

Andrea L. Bonnicksen • Department of Political Science, Northern Illinois University, DeKalb, Illinois

David W. Drebushenko • Department of Philosophy, Central Michigan University, Mount Pleasant, Michigan

Roger B. Dworkin • School of Law, Indiana University, Bloomington, Indiana

Mary B. Mahowald • Pritzker School of Medicine, The University of Chicago, Chicago, Illinois

Christine Overall • Department of Philosophy, Queens University, Kingston, Ontario, Canada

Wade L. Robison • College of Liberal Arts, Rochester Institute of Technology, Rochester, New York

Barbara Katz Rothman • Department of Sociology, Baruch College, CUNY, New York, New York

Thomas A. Shannon • Department of Humanities, Worcester Polytechnic Institute, Worcester, Massachusetts

Bioethics and the Fetus

Introduction

All of the articles in this volume of *Biomedical Ethics Reviews* deal with moral and/or legal problems involving fetal life. Eight articles comprise the volume; summaries of all eight articles follow.

In his chapter, "Common Sense and Common Decency: *Some Thoughts About Maternal–Fetal Conflict*," Roger Dworkin examines some of the basic moral and legal conflicts between the right of the fetus *in utero* and the rights of the woman carrying the fetus. Must pregnant women, for example, be legally constrained to avoid employment in environments exposed to mutagens or taratogens? Must they avoid consumption of licit and illicit drugs including alcohol, tobacco, and caffeine? Must they make "wise" decisions about childbearing and allow medically recommended intrusive procedures to benefit their fetuses, or submit to caesarean deliveries to maximize the chance for good health of the fetus? Such issues have led to a number of recent legal and moral disputes.

Dworkin reviews and discusses various legal cases in which the rights of the fetus conflict with the rights of the mother and basically argues that most cases can be resolved by appeal to common sense and decency. He takes exception to *Roe* v. *Wade* as establishing the status of the fetus in fetal/maternal conflicts. He also takes exception to any other legal precedent for resolving such conflicts. In the course of his discussion, he raises the question of whether legislators should criminalize cocaine use by pregnant women or criminalize the taking of drugs that are already illicit. Should we make it a crime for a pregnant woman to knowingly use cocaine? His answer is that criminalizing such

From: *Biomedical Ethics Reviews • 1991*
Eds.: J. Humber & R. Almeder ©1991 The Humana Press Inc., Totowa, NJ

behavior will probably have more bad effects than good. Should we take the child legally after he or she is born? Should we authorize employee fetal protection plans? Should we legally require medical intrusions to benefit the fetus? Should we legally force, for example, women with excessive phenylaline (causing PKU) in the blood to take a low protein diet to save the fetus from severe mental retardation? In seeking to answer these questions, he examines a number of legal cases.

His general position on the question of whether we should impose medical procedures to protect the fetus is "sometimes." Here again, though, common sense and decency are the guiding lights.

In "Biological Mothers and the Disposition of Fetuses After Abortion," Christine Overall examines the ethical and legal issues affecting fetuses after abortion. She begins by asking whether it is morally acceptable to preserve, against the woman's wishes, fetuses that survive or can survive abortion. Should a pregnant woman be entitled to be assured of the fetus' death either during or after the abortion? Who has the right to determine what to do with the fetus? Does the right to abortion carry with it the right to terminate the life of the abortus? These questions raise the central problem of the maternal–fetal relationship, reproductive autonomy, control over the body, and the status of the fetus.

Overall examines four basic arguments asserting that the failure to observe the wishes of the biological mother with respect to the death of the fetus is morally wrong. In the end, she argues that although nobody has more of a right to decide the disposition of the abortus than the pregnant woman, that does not imply that the choice of death is *eo ipso* morally correct. The pregnant woman does not have a moral entitlement to the death of the fetus if it survives the abortion. Even so, there is no case wherein her decision will be unacceptable.

Davis v. *Davis* is a divorce case in which both parties seek custody of seven frozen embryos that have been created in vitro by Mr. Davis' sperm fertilizing Mrs. Davis' eggs. The court's first step in deciding the issue of custody is to claim that the embryos should be viewed as "children in vitro" rather than as mere property. Having made this determination, the court then asserts that the embryonic "children" have an interest in being born, and that temporary custody of these children should be granted to Mrs. Davis "to assure their opportunity for a live birth."

In "Frozen Embryos and Frozen Concepts," Wade Robison examines the court's opinion in *Davis* v. *Davis*. Robison claims that the court's classification of the frozen embryos as children in vitro is not morally innocent because it assumes that the embryos have moral standing, and so tacitly commits the court on a number of issues in normative ethical theory. As an alternative to classifying the embryos as children, Robison argues that we should treat them in the same way as we treat those who are in a persistent vegetative state, but who are not brain dead. On this view, the embryos would be accorded respect because of the fact that they were genetically human, but they would have no standing in our moral world. Also, the issue of the embryos' disposition would be settled by focusing on the competing interests of Mr. and Mrs. Davis, and not by appealing to the putative interests of the embryos themselves.

In 1988, Abe and Mary Ayala were told that their daughter, Anissa, had chronic myelogenous leukemia, and that the only hope for a cure was a bone marrow transplant. When no suitable donors could be found, Mary Ayala conceived a child with the express intention of using that infant's bone marrow cells to save Anissa's life. A number of ethicists have found fault with the Ayalas' actions, arguing that it violates the Kantian dictum that we should always treat humanity as an end, and never simply as a means. In "Creating Children to Save Siblings' Lives: *A Case Study for Kantian Ethics,*" David Drebushenko takes issue with this claim. He argues that there is some question whether Kant intended for his dictum to apply to actions affecting newborn infants, and that even if this was Kant's intention, a close reading of Kant's moral imperative indicates that the Ayalas' actions do not violate it, but rather accord with its dictates.

In "Fetal Tissue Transplantation: *An Update,*" Mary Mahowald discusses the moral issues attending advances in techniques of fetal tissue transplantation, and an argument provided thereby for opponents to legalized abortion. She examines closely two aspects of fetal tissue transplantation that have emerged as central to the debate: the use of neural tissue and the association between abortion and fetal tissue transplantation. In doing all this, she discusses different frameworks for determining the ethics of fetal tissue transplantation.

One of the basic issues discussed is whether elective abortion for the purpose of providing for tissue transplant should be allowed, or rather, should we use only ectopic pregnancies as a source of fetal tissue transplant? The surgical removal of an ectopic pregnancy is "comparable to the therapeutic abortion for maternal health, with the added caveat that the circumstances are already fatal for the fetus." Even so, on therapeutic grounds, Mahowald and others favor, for various reasons, elective abortion as a source of fetal tissue. As she notes, though, that in itself may not constitute sufficient moral justification for the practice. This is because if the endorsement of the procedure leads to widespread increase in elective abortions, a reduced sense of the value of human life, and to exploitation of women, it is possible that such an array of undesirable consequences would outweigh the potential benefit of the technique. She also discusses the morality of the issue from the deontological perspective. Others oppose fetal tissue transplantation because it endorses the institution of abortion as a preferred supplier and is directly involved with furthering the evil of abortion.

She further examines the moral issues involved in fetal transplantation when the issue depends on transplantation from living donors, or cadaver donors, or from surrogate motherhood. In the end, she concludes that different moral frameworks provide different moral answers. She offers no argument for the supremacy of one moral framework over the other.

In "The Moral Significance of Brain Integration in the Fetus," Thomas Shannon contends that the human fetus has a developing moral standing as it moves into each new stage of its biological evolution; he then seeks to determine the importance of the various stages of brain formation for this developing moral standing. The first step Shannon takes in his investigation is to examine the biological data regarding fetal brain development. Next, Shannon critically evaluates five alternative views concerning the moral significance of these data, and draws several conclusions. The first conclusion Shannon draws is that the concept of brain life, as a mirror image of that of brain death, can be useful in helping to determine the moral standing of the fetus at various stages in its development. In this connection, Shannon stresses the moral significance of the first presence of fetal neural activity at eight weeks prenatal development, and the importance of the integration of the fetus' entire nervous system at twenty weeks gestation. Shannon claims that although fetuses at both of these stages of development are worthy of some degree of respect and protection, neither has the moral status of an actual person, and fetuses of twenty week gestation are worthy of more respect than eight-week-old embryos.

In "The Embryo as Patient: *New Techniques, New Dilemmas*," Andrea Bonnicksen reviews the current literature and identifies the ethical issues attending our recent advances in reproductive technologies. Specifically, she reviews emerging techniques in embryo micromanipulation and identifies those ethical issues that we must examine before we systematically offer those techniques in medical clinics.

After describing the techniques associated with embryo biopsy, embryo microsurgery, and genetic therapy, Bonnicksen reviews basic questions on informed consent, truth-telling, and confidentiality. She also asks the basic questions of why we do these things, and whether we should do these things. She also examines what the basic future ethical issues in this area will be like.

Typical case studies in biomedical ethics encourage students to view ethical decision-making as an individual affair. Further, women who have had prenatal diagnosis and selective abortion tend to share this view, for they see themselves as solely responsible for their decisions to abort. In "Prenatal Diagnosis," Barbara Katz Rothman claims that women's reproductive choices are not wholly free, for they are limited by social, political, and economic conditions that mere individuals are powerless to change. Given this view, Rothman contends that ethical dilemmas are socially constructed, and that women who undergo selective abortions should not see themselves as villains aborting "inconvenient" fetuses, but rather as "victims of a social system that fails to take collective responsibility for the needs of its members, and leaves individual women to make impossible choices."

Common Sense and Common Decency

Some Thoughts About Maternal–Fetal Conflict

Roger B. Dworkin

Introduction

Every silver lining has a cloud.* Progress in the medical and biological sciences has improved prospects for the birth of healthy children and for the well-being of the women who carry and give birth to them. Increasing knowledge about fetal development combines with new knowledge about teratogens and mutagens, the genetic information explosion, and the growth of prenatal diagnosis to increase the chance that babies will be born healthy and that women will be able to make informed choices about childbearing. Developments in prenatal surgery and other medical interventions contribute, and promise to contribute a great deal more, toward the goal of healthy babies. Increasing safety of

*Is this my line? I think so, and I cannot find it in the standard books of quotations, but it seems like the kind of thing somebody would have said before. If the line is yours, I apologize. I did not mean to steal it.

From: *Biomedical Ethics Reviews • 1991*
Eds.: J. Humber & R. Almeder ©1991 The Humana Press Inc., Totowa, NJ

caesarean section deliveries offers health benefits to mothers as well as their offspring.

Yet these advances also generate new conflicts. If we can identify teratogens and mutagens, then must pregnant women or women of childbearing age avoid them? That is, may their employment opportunities be constrained, and must they change their behavior regarding consumption of licit and illicit drugs, alcohol, tobacco, and caffeine, and avoid daily activities as prosaic as changing the kitty litter? Must they make "wise" decisions about childbearing, allow bodily intrusions to benefit their fetuses, and submit to caesarean deliveries to maximize the chances for fetal life and good health?

Lest anyone think these conflicts are not real, they have already given rise to legal disputes, which we shall examine below, and to a growing literature[1] whose language makes the depth of feeling obvious. For example, Annas has written that to treat a fetus against its mother's will "requires us to degrade and dehumanize the mother and treat her as an inert container;" to favor the fetus over the mother in the fetal abuse area "radically devalues the pregnant woman and treats her like an inert incubator or a culture medium for the fetus."[2]

Seeing things somewhat differently, George Smith writes, "[C]ourts should act (with or without personal agreement or acquiescence) to prevent continued tragedies of birth where egregious cases of maternal negligence or culpable behavior have clearly shown that a woman is not deserving of the dignity and moral recognition of a true mother."[3]

Adopting either of these viewpoints leads one to clear results in cases of maternal–fetal conflict, but one can hardly expect either of them to persuade the unpersuaded. Moreover, extreme responses ignore the fundamental insight that law "mediates most significantly between right and right."[4] Much that is "right" or good exists on both sides of every maternal–fetal conflict. The job of the law is not to seize on one "right," thereby converting its opposite number into a "wrong."[5] Rather, sound policy requires accommodation in an attempt to sacrifice as little as possible of what is "right" about each position.

Different kinds of maternal–fetal conflicts can arise in different legal settings. Careful analysis of each conflict in each setting is more likely than across-the-board ideological decision-making to lead to sound results. Surprisingly (or perhaps not?), a common sense devotion to practicality, a commitment to simple decency, and a steadfast determination to minimize the costs of mistakes are effective guides to the resolution of conflicts between pregnant women and the unborn.

Basically, two types of maternal–fetal conflict arise: In one the fetus will benefit from some intrusion of the mother's body. Prenatal surgery or medical treatment is indicated, or the fetus' chance for survival and good health suggests a caesarean delivery. If the mother refuses the procedure, a conflict exists regardless of the reason for her refusal. In the second situation the mother's behavior, independent of any alleged need for medical or surgical intervention, may injure the fetus. The mother may work or seek employment in a dangerous environment, drive negligently, use drugs, drink alcoholic or caffeinated beverages, and so on. If she cannot or will not alter her behavior, her interests and those of her fetus conflict.

The law can confront these conflicts in different ways. It can attempt to criminalize maternal behavior that injures or threatens to injure the mother's fetus. It can attempt to terminate the mother's parental rights or remove custody from her after the child is born. It can enforce private employers' plans to exclude pregnant women or women of childbearing age from certain workplaces. It can give the fetus or the child after birth a civil action against the mother. Or it can try to force the mother to accept an undesired medical or surgical procedure or other behavior change.

Nothing in the law suggests any need to reach one conclusion to apply in all of these instances. Certainly, *Roe* v. *Wade*[6] and its progeny do not clearly force us one way or the other.

An advocate of maternal rights might argue that *Roe* v. *Wade* requires a preference for the mother: *Roe* recognized a constitutional right of privacy that is "broad enough" to include the decision to have an abortion, thus ending the fetus' existence. This right emanates from the Supreme Court's commitment to the value of personal autonomy and its unwillingness to force women into undesired, unpleasant, and stressful situations. The state's interest in the potential life of the fetus does not become compelling until viability. After viability, the state may regulate and even prohibit abortions to preserve fetal life, but even then some interests of the mother must prevail.

Fetal rights advocates, on the other hand, can also find support in *Roe* and the cases following it. *Roe* v. *Wade* expressly rejected the idea of a constitutional right to do what one wants with one's body and authorized states to prohibit persons other than physicians from performing abortions. It also authorized state regulation of abortion from the end of the first trimester in order to protect the state's compelling interest in maternal health. In *Webster* v *Reproductive Health Services,*[7] decided in 1989, Justice Blackmun wrote a dissenting opinion in which, quoting from Justice Stevens, he articulated on behalf of himself and two other members of the Court as strong an autonomy-based position as one is likely to see: "It is this general principle, the "'moral fact that a person belongs to himself and not to others nor to society as a whole'"...that is found in the Constitution."[8] Yet, not only is this a minority position, but it seems not truly to represent even the minority's view because, in the same paragraph, Justice Blackmun reconfirmed his commitment to the view that the state's interest in protecting the health of pregnant women is a legitimate counterweight to the woman's autonomy.[9]

Moreover, every Supreme Court Justice has recognized since 1973 the state's compelling interest in the potential life of the fetus. At least three (Rehnquist, White, and Kennedy)[10] and probably five (O'Connor[11] and Scalia[12]) Justices now believe that interest is compelling throughout pregnancy. All nine Justices authorize states to require activities *before* viability to serve the state's interest that becomes compelling at viability.[13] Finally, the only maternal interests that must prevail over the state's interest in postviability potential life are the mother's interests in her life and health.

At bottom neither of these arguments is very persuasive. *Roe* and its progeny were about abortion. Maternal–fetal conflict is about the obligations, if any, that a pregnant woman who does not get an abortion bears to her fetus.[14] Obviously, the fact that a woman may lawfully prevent her fetus from living does not compel the conclusion that she may inflict any sort of damage she wants on it. The right to turn off a terminally ill patient's respirator does not include the right to stab the patient repeatedly about the face and head.

The fact that neither *Roe* nor any other legal source compels one conclusion to all maternal–fetal conflict problems leaves us free to analyze each problem independently.

Criminalization

Some prosecutors and legislators are attracted to the idea either of convicting pregnant women who use illicit substances of an existing crime, like furnishing drugs to a minor, or of enacting new criminal statutes directed specifically at a pregnant woman's behavior. Ultimately, efforts to use existing laws should fail because of the well established position that criminal statutes are to be construed narrowly in favor of the accused[15] and its constitutional counterpart that renders void any criminal statute that is so vague that it fails to give fair notice to potential violators or adequately to restrict the discretion of law enforcement personnel.[16]

The more interesting question is whether legislatures should enact new statutes to criminalize the use by pregnant women of already illicit substances. For example, should a sensible legislator vote for a bill to make it a crime for a woman who knows she is pregnant to knowingly use cocaine?

The best analysis of when the criminal sanction is a sensible legal response to undesirable behavior was developed by Packer

in his classic *The Limits of the Criminal Sanction.*[17] Beginning with
the recognition that criminalizing behavior is the most
extreme response our legal system has to individual behavior and
that use of the criminal sanction is morally questionable because
it involves the state in intentionally inflicting suffering on a human
being, Packer argues for restricting the use of the criminal law.
Applying his analysis leads to the conclusion that even knowing
cocaine use by a pregnant woman ought not to be a crime greater
than cocaine use by anybody else.

Although a consensus that cocaine use by a pregnant woman
is immoral may exist, criminalizing such behavior will probably
have more bad effects than good. It is unlikely to deter behavior
that is undeterred by the combination of concern for one's own and
one's baby's health and existing criminal sanctions. Instead, criminalization will probably have the perverse effect of increasing
the number of "crack babies" by frightening pregnant women
away from drug abuse treatment and counseling programs. It will
also increase other afflictions of newborns by deterring women
from seeking prenatal care.

Criminal penalties for cocaine use by pregnant women will
be completely unenforceable unless they are accompanied by
imposing an obligation on doctors and other health care providers
to report violations by their patients. This obligation will not only
distort the doctor–patient relationship, but will also force those
women who are not frightened away from seeking care to pay a
premium (what Packer calls a "crime tariff") to those physicians
who are willing to violate the law for a fee. Moreover, enforcement will require intrusions into the bodies of mothers and babies
to perform the necessary drug tests. These "searches and sei-
zures" will have to be litigated to ascertain their constitutionality.
This increases the cost to society of adopting an approach that is
not likely to provide much benefit. Moreover, imposing criminal
penalties on pregnant cocaine users will create opportunities for
inappropriate behavior. The law could not be uniformly enforced.
It would invite sporadic enforcement, which is inevitably dis-

criminatory in a sense, and is likely to discriminate seriously against poor women and women of color. Finally, criminalization would provide angry persons, especially angry men (ex-husbands, lovers, pimps) with a wonderful tool for extortion.

Given the small likelihood of good results, the large likelihood of bad ones, and the clear costs involved, criminalization of cocaine use by pregnant women seems ill advised. If that is so, criminalization of otherwise licit behavior (e.g., smoking tobacco) by pregnant women is even more clearly inappropriate. Even if such behavior is easier to deter than the use of illicit substances, the absence of moral consensus about the behavior, enforcement costs, and excessive intrusiveness into women's lives should preclude use of the criminal sanction in these instances.

Custody and Parental Rights

Refusal to criminalize behavior, of course, does not mean the behavior falls outside the zone of legal intervention. A second potential device for dealing with drug-using pregnant women would be to remove custody from the woman or terminate her parental rights *vis a vis* her child after the child is born. This approach may be no more practical than criminalization if deterring inappropriate prenatal behavior is its goal. However, if the goal is to protect the health and welfare of children after they are born, this family law approach is not necessarily doomed to failure.

Once a child is born it becomes a "person...in the whole sense,"[18] entitled to a full range of legal protections. Parents are considered the natural guardians of their children; they have obligations to provide care and support and reciprocal rights to the custody and society of their children. Occasionally, parental rights may be terminated or custody removed.

Termination of parental rights is an extreme sanction. It ends forever the legal relationship between parent and child. Consequently, the standard for termination is very high. Although precise standards vary from state to state, parental rights will generally not be terminated unless a parent is found to be "unfit." Unfitness means that the parent has engaged and is likely to continue to engage in behavior highly detrimental to the child. A mere finding that the child would be better off living with someone else will not suffice.[19]

The termination sanction is so extreme that its use in "prenatal child abuse" cases is probably not warranted. If drug abuse alone were a sufficient ground for deciding that a parent is unfit, millions of American parents would lose their parental rights.

The fact that a woman used drugs during pregnancy adds one argument for finding her unfit: It indicates that she has already hurt the child, perhaps even knowingly. Nonetheless, termination of parental rights is unwarranted. A single instance of child abuse would not normally cause a parent to lose parental rights, and the fact that the abuse occurred before birth, when the child was still a fetus, a being of uncertain legal stature, weakens rather than strengthens the claim for termination. Moreover, termination of parental rights would remove from a woman one major incentive for attempting to stop using drugs. Although children should not be used as "carrots" to entice their parents into good behavior, the "sticks" the law uses should not remove natural incentives to good behavior if doing so can reasonably be avoided. Termination of parental rights because of prenatal drug use seems unwise.

Whether to remove custody from the mother who abused drugs during pregnancy is a more difficult question. An award of custody is a legal placement of a child and assignment of a primary caretaker. A noncustodial parent, however, retains rights and obligations to the child, and custody decrees may be reopened and custody assigned to a previously noncustodial parent.[20] Therefore, removing custody is not as extreme a course of action as terminating parental rights.

Custody disputes typically arise either between a parent and the state or between parents. In general, the standard for assigning custody is the best interests of the child, although if the state is a party, the standard will vary depending on the language of the state's child abuse and neglect laws. The law ordinarily starts from the presumption that a child's best interests require that custody be with a parent rather than the state or a third party. The modern view in custody fights between parents is that no presumption exists as to which parent should have custody.[21]

In an action by the state to remove custody from a mother because she abused drugs during pregnancy, the presumption in favor of parental custody and the possibly grim placement opportunities for the child if made a ward of the state both suggest the mother should retain custody. Moreover, if one were to imagine a mother who began using drugs only after her child was born, the fact of drug use alone would not be sufficient to place custody in the state. Drug use does not preclude the possibility that a woman will be an adequate mother. However, it may raise questions about her future performance and about where the best interests of the child lie. Thus, postnatal drug use seems relevant to the question of custody.

Prenatal drug use that continues after birth is at least equally relevant. Arguably, it is an even stronger indication that the best interests of the child require removing custody from the mother because the mother's use of drugs during pregnancy has already hurt the child and her continuing drug use may reduce the mother's ability to deal with the special problems of a drug-affected infant. In any event, it seems unrealistic to deny the possibility of a negative impact on the baby as indicated by prenatal maternal drug abuse. Moreover, the intrusion into a woman's autonomy involved in holding her accountable for her criminal conduct is not extreme. Thus, evidence of prenatal drug abuse and of whether it is continuing should be relevant to a custody question between the mother and the state, but should not necessarily determine the outcome. Other factors (maternal affection and concern, efforts to get off drugs, alternative placement opportunities, and so on) are also relevant. This is the position taken recently by the

Appellate Division of the New York Supreme Court in *Matter of Stefanel, Tyesha C.*[22] There, the court held that allegations of prenatal cocaine and marijuana use plus the presence of cocaine in the baby's blood stream at birth were sufficient to state a claim for declaring a child neglected and removing custody from the mother. However, the allegations, if proved, were not conclusive on those questions.

In a custody battle between parents, one parent's use of illicit drugs has the same relevancy as in a case brought by the state. The interesting question is whether the father's drug use cancels out the mother's use and makes drug use unavailable as a consideration about what is in the child's best interest. One can imagine a fairly easy case in which one parent occasionally uses small amounts of drugs and the other is a major drug abuser. But suppose maternal and paternal drug use patterns are about the same. The father will argue that his drug use does not show as much about him as a parent as the mother's does. After all, she has already demonstrated her willingness to hurt the child by taking drugs during pregnancy; he has not. If one assumes the mother's free choice in taking drugs, the father's argument has some grounding in fact. However, the father did not hurt the child by taking drugs during pregnancy because doing so is impossible, not because of any virtuous behavior or commitment to the child. This, plus the fact that a similar argument would never be available to a mother, suggests an unfair male bias in weighing a mother's prenatal drug abuse more heavily than a father's drug abuse. The best solution is probably to consider what each parent's behavior at the time of trial suggests about his or her future dealings with the child, i.e., what custodial placement (which is for the future) will be best for the child (in the future).

Fetal Protection Plans

A similar problem is posed in considering employers' so-called fetal protection plans. Here too, biological differences between men and women raise the question of what fair or "equal" treatment requires.

Fetal protection plans are workplace rules that exclude pregnant women (or women who may become pregnant) from jobs in settings that are thought dangerous to their future offspring because workers are exposed to some teratogenic or mutagenic agent. For example, in the *Johnson Controls*[23] case, which is pending before the Supreme Court as this chapter is being written, the employer manufactures batteries. Battery making exposes workers to lead, which, in turn, is said to pose a danger to the unborn children of exposed women.

Whether fetal protection plans are legally permissible is a very important question. Some estimates place the number of jobs that could be closed to women if fetal protection plans were generally adopted as high as 20,000,000.[24] Even if that figure is too high, fetal protection plans obviously prove a very substantial issue for the financial, social, and symbolic status of women.

As *Johnson Controls* will soon be decided, I shall only outline briefly some of the issues that have to be resolved about fetal protection plans.

One question is what motivation underlies such plans. If, for example, an employer's primary goal is to avoid liability rather than to protect fetuses, that goal can be accomplished through a simple change in the law that would render fetal protection plans unnecessary. If the law required employers to make reasonable efforts to learn about and inform workers about fetal hazards, and allowed a parent's consent to the risk to bind an after-born child, employers' liability concerns would be resolved.

If the goal of fetal protection plans is really to protect fetuses, then one may wonder why fetal protection plans do not apply to men. Equal treatment for the sexes suggests itself as a necessary, but not sufficient, criterion for the acceptability of a fetal protection plan. Resistance to such an approach would lend credence to critics who see fetal protection plans as plans for the perpetuation of male domination.[25]

Challenges to fetal protection plans so far have argued that the plans discriminate against women in violation of Title VII of the 1964 Civil Rights Act.[26] In order to evaluate conduct to decide whether it violates Title VII, one's first question must be whether the conduct (here, adoption of a fetal protection plan) intentionally discriminates against a protected group (women) or whether an unintended, but nonetheless real, disparate impact falls adversely on women.[27] If discrimination is intentional, the plan will be invalid unless it can be justified as a bona fide occupational qualification (BFOQ).[28] A BFOQ exists if only members of one sex could do the job. Almost no such jobs exist. The practical consequence of finding intentional discrimination will be to invalidate a fetal protection plan.

If the plan does not intentionally discriminate, but has an adverse disparate impact on women, it can be justified by the employers' proving a business necessity for the plan.[29] This is an easier showing for an employer to make, but it still requires the employer to validate the information on which its policy is based and to show a legitimate business reason for the policy.

Although crystal ball gazing is perilous, a likely outcome of the fetal protection plan issue would be that some plans will be upheld, and that the acceptability of each plan will have to be decided on a case-by-case basis. Presumably, the minimum requirements for an acceptable plan should be

1. A severe threat to fetuses;
2. Based on highly probative and reliable data; and
3. A narrowly drawn plan that closes the minimum possible number of jobs to women.

Additional desirable requirements would be insistence that a plan protect affected women's pay, benefits, seniority, authority, and status when possible, and that the employer have similar programs it applies to men whose jobs place their offspring at similar potential risk. These last two suggested requirements are important. Protection for women's authority and status as well as pay, benefits, and seniority recognizes the nonfinancial as well as the financial rewards of employment. It makes fetal protection plans very expensive, thereby discouraging them without prohibiting them. Requiring that women be protected when possible (not when reasonably possible) means that women affected by fetal protection plans should always be protected unless the employer has no safe jobs, and, perhaps, when protecting the woman would require firing an existing employee (who may be either a woman or a man).

The protection of existing positions does nothing to protect women who are not hired because of fetal protection plans. The requirement of equal treatment for men, however, would protect women as a group, although not a particular woman (or a particular man) from differential treatment. The requirement of equal treatment of men would impose an enormous burden on employers to amass data and incur financial costs. Again, this should discourage fetal protection plans without prohibiting them and assure that any acceptable plan really is for the protection of fetuses.

This approach, which uses the law to influence choices but leaves the power to choose to individuals, is a very attractive way to mediate between competing values. In the fetal protection plan context it will lead to minimal limitations on women's opportunities and will provide exactly as much fetal protection as we as a society really want (i.e., as much as we are willing to pay for).

Finally, fairness to employers requires that an employer not be held liable for failure to implement an invalidated fetal protection plan and not be obligated to develop a fetal protection plan. Of course, an employer may be liable for failure to take other reasonable measures to protect fetuses (e.g., making reasonable efforts to render the workplace safe) and for failure to inform and warn employees of danger to their offspring. As suggested before, a properly warned employee's consent to run the risk should bar later born children from recovering against the employer.

Supervision of Daily Life

Even more troubling than the issues discussed so far are maternal/fetal conflicts that raise the spectre either of supervision of the ordinary details of a woman's life or the imposition of medical or surgical intrusions upon her. The detail and intimacy of these intrusions combines with the likelihood that they will be the most important to the well-being of the fetus to paint the conflict between mother and unborn child in sharpest relief. Here too, however, the search for practical answers that do not violate basic norms of decency and that minimize the costs of mistakes by encouraging desired choices rather than resorting to compulsion offers hope for sound results.

Can a woman be compelled to undergo a medical regimen for the benefit of her fetus? Phenylketonuria (PKU) is an autosomal recessive disease in which the body cannot properly metabolize phenylalinine (an essential protein). The excess phenylalinine damages the brain and causes severe mental retardation and behavioral difficulties. If the condition is diagnosed in the first few months of life, and the infant is placed on a highly restrictive low phenylalinine diet, the symptoms can largely be avoided. Since the mid 1960s, newborn screening for PKU has been the norm in the United States.[30] The result is that treated persons are now functioning normally in the community.

Unfortunately, this silver lining is like all others. When women with PKU lived in institutions, they seldom reproduced. Now, healthy women with PKU do become pregnant. When a woman who has PKU becomes pregnant, the excess phenylalinine in her blood will cross the placenta and severely damage the fetus' brain. The child will not have PKU, but will be severely mentally retarded.

Some evidence indicates that if a woman with PKU resumes the low phenylalinine diet during pregnancy (and ideally for a few months before pregnancy) the harm can be avoided. Should such a woman be forced to resume the PKU diet?[31]

The diet is very restricted and unpleasant. In essence, it requires the dieter to avoid all protein sources and to utilize an unpalatable low or no phenylalinine product as the dietary staple. This is not like asking someone to watch calories, fats, or cholesterol.

Although a woman probably has no "right" to refuse to go onto the diet, a decision by the law to try to force her to do so would be singularly ill-advised.

A woman could obviously be restrained from bashing in the brains of her newborn child. Presumably, she could also be restrained from injecting a brain destroying chemical into the amniotic sac the day before her scheduled delivery date. Why then would anyone think she has a right to slowly destroy the fetus' brain by sending a brain destroying substance across the placenta? The suggestions that the phenylalinine is naturally in her body and that her refusal to diet is a mere omission are unpersuasive. The law often compels persons to act when a good enough reason to do so exists, and the naturalness argument smacks too much of picking and choosing which natural phenomena to insist on. After all, the state required the woman to be screened in infancy, and only her treatment by the unnatural, low phenylalinine diet has allowed her to advance to the point at which she is uninstitutionalized, pregnant, and faced with a choice to make.

The real reasons not to force the woman to resume the low phenylalinine diet are that doing so is impractical and indecent. A woman cannot be forced onto the diet unless she is watched and supervised every second of every day throughout her pregnancy. She must be force-fed the diet if she refuses to eat and physically restrained from eating any unpermitted food. The resources required for such an effort would be enormous and hardly the best use of limited health-care dollars in a nation beset by high infant mortality, a low level of prenatal care, malnutrition, AIDS, inadequate medical insurance, and so on, and so on. And what kind of a system would take a woman whose only sins are pregnancy and stubbornness and confine her, force-feed her, restrain her, and goodness knows what else? In a country whose judicial conscience is shocked by pumping the stomach of a suspected criminal to get him to disgorge narcotics,[32] this simply will not do.

Civil Liability

Suppose then, that the properly informed woman refuses to follow the diet and gives birth to a brain damaged child. May the child recover damages from the mother in a civil action? Certainly legal doctrine exists that could be applied to support the child's claim. The important question is whether choosing to apply the doctrine in that way would represent sound social policy.

Doctrinally, the basis for recovery for personal injuries not caused by defective products or extraordinarily dangerous activities (like blasting) is negligence. That is, an injured person must prove that the defendant behaved negligently and that the negligent behavior caused the injury. In the ordinary case liability follows from a finding of negligently caused injury. Occasionally, however, some reason of social policy will exempt a negligent defendant from liability for injuries he or she negligently caused. When that occurs, courts say the defendant owed no duty to the plaintiff. To say that a defendant does not owe a plaintiff a duty means that the defendant will not be liable to the plaintiff for injuries his or her negligence caused to the plaintiff. Conversely, to say that a defendant does owe the plaintiff a duty means that the defendant will be liable for injuries he or she *negligently* caused the plaintiff. A duty is never an obligation to accomplish a certain end (rescue a person, cure a patient, and so on). A duty is merely an obligation to behave reasonably (i.e., nonnegligently) toward the plaintiff. Thus, the two critical doctrinal questions in a negligence case brought by a child against its mother to recover for prenatal harm are whether a pregnant woman has an obligation to behave reasonably toward her fetus, and, if so, whether she did behave reasonably toward her fetus.

No sound reason exists to deny a woman's duty to treat her fetus reasonably. The standard litany of reasons to reject such a duty are set out by the Illinois Supreme Court in *Stallman* v. *Youngquist*.[33] There, a child alleged that she was injured by the negligent driving of her mother while the mother was five-months pregnant with the child. The court rejected the child's claim. It said that to recognize that a pregnant woman owes a duty to her fetus would affect the way society views women and their reproductive abilities; it would require a woman to create the best prenatal environment possible; it would make her a guarantor of the fetus' mental and physical health; it would make the mother and fetus adversaries; it would create an impossible situation in which courts could not create a standard against which to measure the woman's performance; and it would be an excessive infringement on women's lives—an unacceptable intrusion on their privacy and autonomy.

If any of that were correct, imposing a duty on pregnant women would be unattractive indeed. However, rhetoric does not make reality, and the court was simply mistaken in its assertions. It made the mistake the unfortunate word "duty" invites. It forgot that a legal duty is only an obligation to act reasonably. Imposing a duty would not require a woman to guarantee the fetus' mental and physical health. It would only require her to treat her fetus reasonably well, not perfectly. The standard of measurement is not problematic; a woman could be held to the same negligence (unreasonableness) standard that every other defendant (car driver, home owner, industry, physician, and so on) has been held to at least since 1850.[34] Courts know how to do a negligence analysis, and the negligence standard is what will protect women from excessive intrusions into their privacy and autonomy, and quash expectations that they must create the best possible prenatal environment.

Perfectly conventional negligence analysis weighs the likelihood and severity of the harm to the plaintiff against such factors as the value of the defendant's behavior, the burden on the defendant of asking him or her to avoid the risk, and the feasibility of behaving differently than he or she did.[35] The burden of placing an unwilling woman onto the PKU diet is enormous. The burden of asking a pregnant woman to drive her car nonnegligently is minute. It asks nothing of her she did not already have to do. Recognizing that pregnant women owe a duty to their fetuses would allow the Stallman baby to be compensated for her injuries, but would almost surely not provide recovery for the injured offspring of the woman with PKU. Moreover, the feasibility inquiry will protect women who do not know they are pregnant from being found negligent towards their fetuses, except in the occasional bizarre case in which a woman unreasonably remains blind to obvious facts well into the fourth or fifth month of her pregnancy. Even there, the law would offer special protection if the woman's lack of awareness could be traced to her youth[36] and maybe even if it could be traced to some recognizable mental deficiency.[37]

The doctrinal analysis shows that the framework for recognizing a pregnant woman's duty to her fetus is present and that neither intrusiveness nor difficulty of setting standards is a valid reason to reject such a duty. Other reasons may be offered, however, to refuse to impose a duty. They too are unpersuasive.

Those who prize stability could argue that recognizing a duty here breaks too much new ground. Taken too seriously such an argument is simply an argument for legal calcification. In this case, moreover, it is not even correct.

Courts long followed the doctrine of intrafamily immunity. That concept precluded close family members from recovering against each other in order to promote family harmony and avoid collusive actions against insurance companies.

The doctrine has now been generally rejected throughout the United States.[38] The existence of insurance reduces the likelihood that lawsuits will create family dissension. Moreover, compensation of injured persons and spreading the costs of injuries now seem sufficient reasons to run whatever risk of dissension remains. Collusive lawsuits designed to cheat insurance companies can be avoided by the usual requirements of medical evidence to support injury claims, cross examination of parties, treating the insured as a "hostile witness," and other judicial devices for ascertaining the truth.

In any event, the concerns about harmony and collusion are no greater in the case of a newborn injured before birth than in the case of a living child. The rejection of intrafamily immunity clears one stumbling block from the potential plaintiff's path.

Similarly, children injured before birth are now routinely permitted to recover against third parties for negligence.[39] Thus, recognizing the "standing" of a child to recover for prenatal injuries requires no extension of existing law. Recovery for prenatal injuries combines with the demise of parental immunity to make the step toward recovery for prenatal injury by a parent a small one for the law to take.

The remaining reason to refuse to impose a tort duty on a woman toward her unborn child is that doing so would, in essence, require the woman to be a Good Samaritan in contravention of the standard American rule that individuals have no obligation to help others.[40] This rule is more honored in the breach than the observance. When the reasons for the rule do not apply, the courts do not apply the rule.[41] As argued earlier, the reasons to refrain from imposing a duty do not apply in the maternal–fetal case.

The relationship between parent and child imposes a duty to behave reasonably, even a duty to act as a Good Samaritan, to one's children.[42] For example, a father or a mother will be liable for failing to seek medical attention for an obviously ill child. Yet neither parent would be liable for failing to donate a kidney to their child who was dying of renal failure. The parent–child relationship, and hence the duty, are the same. The difference in outcome arises from the fact that submitting to surgery and surrendering a kidney are simply too much to ask of a person. The burden on a person from such a situation would be unacceptably heavy. In other words, failure to donate the kidney is not negligent. All parents owe a duty to behave nonnegligently toward their living children. The negligence requirement keeps liability in bounds.

The same situation can and should obtain prenatally. Suppose a married man knowing that he has syphilis keeps that information from his pregnant wife and has intercourse with her, thus causing her to contract syphilis, which she passes to her infant. No sound reason of policy would exempt the man from liability to his child. He behaved unreasonably toward the child before birth, and injured it. Liability is not burdensome, difficult to administer, unlimitable, unjust, or in any way inappropriate.

Similarly, if a woman behaves negligently toward her fetus and injures it, she too should be liable. The difficulty of proving negligence will result in very few cases of liability, but the small gain will be worth its small price. Women, like men, will be accountable for their conduct. Case by case adjudication, sensitive to the facts of each situation, will maximize the likelihood of achieving sound results in particular cases. No woman will be forced to do anything, but the law may affect some women's choices by telling them what society expects and leaving open the possibility of liability. In a highly charged area, with much that is good on both the woman's and the fetus' sides of the ledger, that seems about as good a job as a legal system can do.

Caesarean Sections

More troubling than imposing liability on a pregnant woman for negligent behavior toward her fetus is the possibility of compelling the woman to undergo a medical procedure for the fetus' benefit.[43] Many cases have compelled women who are Jehovah's Witnesses to accept blood transfusions either to keep the women alive to be mothers to their already born children or to improve a fetus' chance of survival.[44] Cases attempting to force women to undergo intrauterine surgery for the benefit of their fetuses are easy to envision. For the moment, however, the paradigmatic case of compulsory medical intervention on a woman to benefit a fetus is the effort to compel a woman to undergo caesarean section delivery. Although a number of such efforts have been made, only two reported appellate decisions exist, one from Georgia, which compelled the caesarean procedure, and one from the District of Columbia, which refused to do so.

In *Jefferson* v. *Griffin Spalding County Hospital Authority*,[45] a competent woman with placenta previa refused a C-section delivery, but was forced to have it anyway to benefit both her fetus and herself. The District of Columbia case, *In re A.C.*,[46] distinguished *Jefferson* on the grounds that no clear maternal/ fetal conflict existed in *A.C.* and that the caesarean in *A.C.* was only for the benefit of the fetus. Despite these distinctions, *A.C.* plainly rejected the acceptability of compelled caesarean sections that *Jefferson* tolerated.

A.C. involved a twenty-eight-year-old married woman, who had had cancer since age thirteen. She married at age 27, while her cancer was in remission, and soon became pregnant. She attended the high-risk pregnancy clinic at George Washington University Hospital. Unfortunately, when she was twenty-five weeks pregnant, she was hospitalized with an inoperable tumor of the lung. Her doctors told her that her condition was terminal. She consented to palliative treatment to extend her life until she was twenty-eight weeks pregnant, at which time she contemplated a caesarean delivery for the child. She consented in advance to that C-section, knowing that the chances of survival for the baby were much greater at twenty-eight weeks than at twenty-six, but also knowing that the palliative treatment "presented some increased risk to the fetus."[47]

After this, chaos ensued. The woman's condition became worse; death was imminent. She drifted into and out of consciousness and expressed different positions about whether she would consent to a C-section at twenty-six-and-a-half weeks. That C-section might shorten her life. It would improve the fetus' chance to survive, however, because the mother could not live until twenty-eight weeks, and the fetus was unlikely to be born alive after the mother's death.

After a hurried, in-hospital hearing, the trial court authorized the caesarean delivery. The appellate court denied a stay.[48] The caesarean was performed. The child died within 2 ¹/₂ hours, the mother in two days. Subsequently, the appellate court considered the merits of the case, despite the fact that time and circumstances had rendered it moot. It reversed the order that had authorized the caesarean.

A.C. involved a patient of questionable competence whose views with regard to the proposed C-section were unclear. Nonetheless, the court began its analysis by considering the issue of compelling a competent woman to undergo a caesarean. This was the court's first mistake.

Courts often decide medical cases involving incompetent persons by first deciding what the rights of a competent person would be and then purporting to extend the same rights to incompetents. This is said to recognize the fundamental dignity of persons with handicaps by giving them the same benefits the law gives nonhandicapped persons. This is utter nonsense. As more than one distinguished jurist has noted,[49] treating incompetent persons as if they were competent denies reality, requires the law to overlook what may be the dominant fact of the person's life, compels resort to legal fictions, and applies doctrines where they make no sense. More importantly, far from recognizing the dignity of the incompetent person, this approach demeans that person by ignoring her individuality and treating her not as who she is but rather as the court's idealized version of who she ought to be. This approach results in imposing on an incompetent person the choices that a competent court thinks the incompetent person would make if she could choose. Although this arrogant pretense may make a court feel good about what it is doing, it no more accords dignity to a real human being who has a handicap than calling an orange an apple turns orange juice into apple cider.

Be that as it may, the court did consider the situation of the competent woman who refuses a caesarean and decided that in virtually all such cases the woman's decision controls.[50] This is plainly the result compelled by common sense and decency. The only way to force an unwilling competent woman to undergo a caesarean section delivery would be to seize her and resort to physical violence, tying her to the operating table or anesthetizing her. Not only are fundamental concepts of human decency offended by the spectre of such behavior by the state, but if women seriously believed that such behavior could result from their medical choices, many of them would choose to avoid physicians altogether. The resulting reduction in prenatal care would result in increased injuries and deaths of both mothers and fetuses/children. Thus, one need not resolve the tension between fetal and maternal interests to conclude that compulsory C-sections on competent women are unwarranted. Such procedures are bad for both mothers and fetuses.

Unfortunately, the court in *A.C.* reduced consideration of these practicalities to a footnote[51] and unnecessarily set out to resolve the maternal–fetal conflict question in the case of competent mothers as a matter of principle.

A compulsory caesarean delivery could be viewed as either unwarranted medical treatment for the mother or use of the mother to benefit the fetus. Either way, the court said, it is virtually never acceptable. A woman has a right to accept or reject medical treatment. This right, rooted in the common law and (the court says) in the Constitution, serves the value of personal autonomy in decision-making, i.e., freedom of choice.

One may ask why the mother's autonomy interest should prevail over the fetus' similar interest. Why isn't a one-time intrusion on the mother's autonomy outweighed by the chance to provide the fetus with a lifetime of autonomy? The court's answer is to note that a person has no obligation to sacrifice her body for another living person, and that "surely" a fetus cannot have more rights than an already living person.[52]

With all respect, that answer begs the question in two ways. First, it assumes that the relevant question is fetal rights rather than maternal duties. A mother may have a duty to the state without the fetus acquiring any rights at all. Second, and more important, if the issue is rights, then the question is whether the fetus *should* have more rights than a living person. One could argue that a fetus should have more rights than a living person because of its abject and total dependence on its mother, its lack of any advocate against its mother, and its total lack of representation in the political process.[53] Judge Belson in his dissenting opinion in *A.C.* makes much of the fetus' dependence on its mother and on the mother having undertaken to care for the fetus as reasons to demand more, not less, of mothers toward fetuses than we do of other persons.[54]

An answer to Judge Belson's argument could note that a pregnant woman already sacrifices more for her fetus than the law ever forces anyone else to sacrifice for another, and that to require her to do more would truly exacerbate rather than alleviate differences caused by biology.

At bottom, a person will be forced to choose which of the views he or she prefers if the person is to resolve the question on principle. Since, however, practical considerations resolve it for us, why would anyone choose to choose between the attractive claims of mother and fetus? Not only does choosing unnecessarily offend those who seem to lose, but also, it commits one to a mode of analysis in other cases that may be unsound. That is what happened in *A.C.*

If incompetent persons are entitled to the same rights as competent ones, and if the right at stake is a right based on autonomy and freedom to choose, then a court must confront the question of how to honor the *right* to choose of those who lack the *ability* to choose.

A.C. purported to honor this right by adopting the substituted judgment approach to decision-making for women whose views cannot be known.[55] Under this approach a court is supposed to make the judgment for a person that she would make for herself if she were able to do so. The standard is subjective; it asks what this woman would do, not what a hypothetical reasonable woman would do.

The problem, of course, is that evidence of what the woman would do if she were competent is seldom available. *A.C.* had extensive experience with doctors and hospitals; she knew she had a high-risk pregnancy and was likely to need a caesarean. Yet even in her case, the evidence of her intention was conflicting and ambiguous. There will be even less evidence of intention when the question of an emergency C-section arises for a woman who has never been competent, a victim of a sudden accident, an unsophisticated patient, or simply a woman who has not discussed possible disaster with doctor, family, and friends. In the real world, the substituted judgment test is a device to make courts feel that they are respecting women's choices while imposing judicial ones. Any doubt on that score was resolved by the *A.C.* court, which noted that if the court cannot decide what the patient would choose, the court may consider "what most persons would likely do in a similar situation."[56] So much for subjectivism and individual dignity!

Judge Belson's concurring and dissenting opinion offers a much more sensible approach. Belson suggests balancing the unborn child's interest in life and the state's interest in protecting life against the mother's interests. Belson requires the balancing court to give "great weight" to the woman's decision. "In a case, however, where the court in the exercise of a substituted judgment has concluded that the patient would probably opt against a caesarean section, the court should vary the weight to be given this factor in proportion to the confidence the court has in the accuracy of its conclusion."[57] What could be more sensible than that? Belson recognizes the importance of protecting a pregnant woman's autonomy *and* the foolishness of blindly following the fiction of an incompetent person's choice wherever it may lead. He tells lower courts to decide one case at a time, considering everything that is relevant, thereby maximizing the likelihood of reaching sound results in particular cases. His approach provides no day-to-day guidance for hospitals, but neither does the majority's approach, which requires courts to do a substituted judgment analysis in each case.

Finally, the practical and decency-based reasons for refusing to compel competent women to undergo caesarean deliveries do not apply when the patient is incompetent and has not expressed her will. The possibility that a court will authorize a caesarean section on an incompetent woman will not keep incompetent women from seeking prenatal care. They will either already be under a doctor's care or be in a position in which decision making, based on predictions of legal consequences is hardly likely. Such women also do not confront the law with "arrest" and the use of force to override their will. By hypothesis, nobody knows what their will is.

Conclusion

The rapidly expanding area of maternal–fetal conflict teaches lessons that are relevant to many other areas where biomedical developments place strains on the law. Broad, principled decisions are unlikely to be satisfactory. Each factual and legal context must be analyzed independently because the impact, significance, and implications of each kind of problem and each legal response are different. Foolish consistency really is the "hobgoblin of little minds."[58] The job of the law is to know when a principle has been pushed as far as it *sensibly* can, not as far as it *logically* can. Easy sloganeering unrelated to reality leads to fictions, self-deception by the legal system, and unsound results. A focus on what is practical, avoidance of the shocking or indecent, and a conscious effort to sacrifice the least amount possible of each of the competing values at stake will come as close to tolerable resolutions of insoluble problems as a legal system designed and run by mortals can hope to come.

In the maternal–fetal conflict area decency, practicality, and minimal sacrifice of competing values suggest that the criminal sanction not be used to penalize pregnant women's prenatal behavior; that competent women not be forced to undergo caesarean section deliveries or alter their behavior for the fetus' sake; that industrial fetal protection plans be rigorously controlled and discouraged; that evidence of prenatal conduct be used cautiously, and not conclusively, in custody and parental rights cases; that children have a cause of action against their parents for prenatal negligence; and that decisions about caesarean sections for incompetent women or women whose intentions cannot be learned be made one case at a time considering all the facts and all the values at stake.

Update

While this chapter was in press, the Supreme Court decided the *Johnson Controls* case. *International Union* v *Johnson Controls, Inc.*, —US—, 111 S. Ct. 1196 (1991). The Court held that a single sex fetal protection plan is sex discrimination, which is forbidden under Title VII unless the employer can show a BFOQ. Only conditions that affect an employee's ability to do the job can be BFOQS. Safety related considerations do not qualify unless they relate to the "essence of the business." Johnson Controls could not establish a BFOQ. Therefore, its fetal protection plan was invalid.

The Court noted that if Title VII bars sex specific fetal protection policies, the employer fully informs the employee of the risk, and the employer has not acted negligently, "The basis for holding an employer liable seems remote at best."

Acknowledgments

I wish to thank Patrick L. Baude, Terry Morehead Dworkin, Lynne Henderson, and Julia Lamber for helpful insights they have provided.

References

[1]R. Trammell (1989) Fetal Rights—A Bibliography. *N. Ill. L. Rev.* **10**, 69–87.

[2]G. Annas (1989) Predicting the Future of Privacy in Pregnancy: How Medical Technology Affects the Legal Rights of Pregnant Women. *Nova L. Rev.* **13**, 329–353.

[3]G. Smith (1989) Fetal Abuse: Culpable Behavior by Pregnant Women or Parental Immunity? *J. Law Health* **3**, 223–235.

[4]P. Freund (1970) Legal Frameworks for Human Experimentation, in *Experimentation with Human Subjects* (P. Freund, ed.), Braziller, NY, p. 105.

[5]*See* C. Schneider (1988) Rights Discourse and Neonatal Euthanasia. *Calif. L. Rev.* **76**, 151–176.

[6]*Roe* v. *Wade,* 410 U.S. 113 (1973).

[7]*Webster* v. *Reproductive Health Services,* —U.S.—, 109 S. Ct. 3040 (1989).

[8]*Id.* at 3073 (Blackmun, J., concurring in part and dissenting in part).

[9]*Id.*

[10]*Id.* at 3057 (opinion of Rehnquist, Ch. J.)

[11]*Id.* at 3063 (O'Connor, J., concurring in part and concurring in the judgment); *see also Akron* v. *Akron Center for Reproductive Health, Inc.,* 462 U.S. 416, 452–475 (1983) (O'Connor, J., dissenting).

[12]109 S. Ct. at 3064–67 (Scalia, J., concurring in part and concurring in the judgment).

[13]*Id.* at 3070–71 (Blackmun, J., concurring in part and dissenting in part).

[14]*In re A.C.,* 573 A.2d 1235, 1245n.9 (D.C. App. 1990); J. Robertson (1989) Reconciling Offspring and Maternal Interests During Pregnancy, in *Reproductive Laws for the 1990s* (S. Cohen and N. Taub, eds.), Humana Press, Clifton, NJ, pp. 259–274; *see generally,* J. Robertson (1982) The Right to Procreate and In Utero Fetal Therapy. *J. L. & Med.* **3**, 333–366.

[15]*See* H. Packer (1968). *The Limits of the Criminal Sanction.* Stanford University Press, Stanford, CA, p. 95

[16]*Id.,* pp. 93–94.

[17]H. Packer, supra, n. 15.

[18]*Roe* v. *Wade,* 410 U.S. 113, 162 (1973).

[19]*See* generally, H. Clark (1988) *The Law of Domestic Relations in the United States,* 2nd. ed., West Publishing Co., St. Paul, MN, pp. 850–905.

[20]*Id.,* pp. 786–849.

[21]*Id.,* pp. 799–827.

[22]*Matter of Stefanel, Tyesha C.,* 157 App. Div.2d 322, 556 N.Y.S.2d 280, leave to appeal granted, 559 N.Y.S.2d 813 (App. Div. 1990).

[23]*International Union, U.A.W.* v. *Johnson Controls, Inc.* 886 F.2d 871 (7th Cir. 1989), cert. granted 110 S.Ct. 1522 (1990).

[24]H. Hoffmann (1990) Fetal Protection Policies: A Method of Safeguarding Fetuses or a Way of Limiting Women in the Workplace. *J. Health Hosp. L.* **23,** 193–207,213,224.

[25]*See id.,* p. 193; J. Stellman (1989) Protective Legislation and Occupational Hazards: Flawed Science and Poor Policies, in *Reproductive Laws for the 1990s* (S. Cohen and N. Taub, eds.), Humana Press, Clifton, NJ, pp. 341–351; Comment (1989) The Fetal Rights Controversy: A Resurfacing of Sex Discrimination in the Guise of Fetal Protection (1989). *U.M.K.C. L. Rev.* **57,** 261–288.

[26]42 U.S.C. § 2000e et. seq. (1982).

[27]Hoffmann, supra n.24, p. 194; Lamber (forthcoming 1991) Mothers and Work: Reproductive Hazards in the Workplace, in *Women on the Job in Europe and the USA in the 1980's: Their Social, Economic, and Cultural Status* (M. Rozbicki, ed.).

[28]*Id.*

[29]*Id.*

[30]L. Andrews (1985) *State Laws and Regulations Governing Newborn Screening.* American Bar Foundation, Chicago, IL.

[31]*See generally,* J. Robertson and J. Schulman (Aug. 1987) Pregnancy and Prenatal Harm to Offspring: The Case of Mothers with PKU. *Hastings Cntr. Rep.* **17,** 23.

[32]*Rochin* v. *California,* 342 U.S. 165 (1952).

[33]*Stallman* v. *Youngquist,* 125 Ill.2d 267, 531 N.E.2d 355 (1988).

[34]*Brown* v. *Kendall,* 60 Mass. (6 Cush.) 292 (1850).

[35]*United States* v. *Carroll Towing Co.,* 159 F.2d 169 (2d. Cir. 1947); H. Terry (1915) Negligence. *Harv. L. Rev.* **29,** 40–54.

[36]*See* W. Keeton, et al. (1984) *Prosser and Keeton on the Law of Torts,* 5th ed. West Publishing Co., St. Paul, MN, pp. 179–182.

[37]Cf., *Breunig* v. *American Family Ins. Co.,* 45 Wis.2d 536, 173 N.W.2d 619 (1970).

[38]*See* W. Keeton, supra n.36, pp. 901–907.
[39]*Id.*, pp. 367–370.
[40]*Id.*, pp. 375–376.
[41]*Id.*, pp. 376–385.
[42]*See id.*, pp. 377, 901–907; J. Robertson (1989) Reconciling Offspring and Maternal Interests During Pregnancy, in *Reproductive Laws for the 1990s* (S. Cohen and N. Taub, eds.), Humana Press, Clifton, NJ, p. 260.
[43]*See* generally, N. Rhoden (1986) The Judge in the Delivery Room: The Emergence of Court-Ordered Caesarians. *Calif. L. Rev.* **74**, 1951–2030.
[44]*E.g.*, *Application of President and Directors of Georgetown College, Inc.*, 331 F.2d 1000 (D.C. Cir.), *cert. den.* 377 U.S. 978 (1964); *Raleigh Fitkin–Paul Morgan Mem. Hosp.* v. *Anderson*, 42 N.J. 421, 201 A.2d 537, *cert. den.* 377 U.S. 985 (1964).
[45]*Jefferson* v. *Griffin Spalding County Hosp. Auth.*, 247 Ga. 86, 274 S.E.2d 457 (1981).
[46]*In re A.C.*, 573 A.2d 1235 (D.C. App. 1990).
[47]*Id.*, p. 1239.
[48]*In re A.C.* 533 A.2d 611, (D.C. App. 1987).
[49]*Cruzan* v. *Director, Missouri Department of Health*, —U.S.—, 110 S. Ct. 2841 (1990) p. 2852; *Conservatorship of Valerie N.*, 40 Cal.3d 143, 174–191, 707 P.2d 760, 781–793, 219 Cal. Rptr. 387, 408–420 (Bird, C.J., dissenting).
[50]*In re A.C.*, 573 A.2d 1235, 1237 (D.C. App. 1990).
[51]*Id.*, p. 1244, n.8.
[52]*Id.*, p. 1244.
[53]Cf., *United States* v. *Carolene Products Co.*, 304 U.S. 144, 152–53, n.4 (1938).
[54]*In re A.C.*, 573 A.2d 1235, 1253, 1256 (D.C. App. 1990) (Belson, J., concurring in part and dissenting in part).
[55]*In re A.C.*, 573 A.2d 1235, 1249–1251 (D.C. App. 1990).
[56]*Id.*, p. 1251.
[57]*Id.*, p. 1258 (Belson, J., concurring in part and dissenting in part.)
[58]R. W. Emerson (1841) *Essays*.

Biological Mothers and the Disposition of Fetuses After Abortion

Christine Overall

Introduction

This chapter begins from abortion but is not about abortion. It examines some aspects of the role of the biological mother with respect to the disposition of fetuses after abortions. Should the biological mother be entitled to determine the disposition of the fetus where that disposition is not in the interests of the fetus itself? In particular, should the mother be entitled to be assured of the fetus' death, either during or after the abortion? Is it wrong to preserve, against the woman's wishes, fetuses that do or can survive abortion?

Whereas the numbers of fetuses that survive abortion are small, the question of their disposition is significant because it brings together some central problems concerning maternal/fetal relationships, reproductive autonomy and control over the body, and the status of the fetus. Before pursuing these, it is necessary to make some remarks about the nature of the question.

First, it might seem inappropriate to use the term "fetus" with respect to an entity that survives abortion. In legal and medical

From: *Biomedical Ethics Reviews • 1991*
Eds.: J. Humber & R. Almeder ©1991 The Humana Press Inc., Totowa, NJ

contexts, "fetus" is usually used to refer only to the "product of conception" *before* it has emerged from its mother's body (Law Reform Commission of Canada, 1989). As a result, the terms "abortus" or "infant" are sometimes used to describe the entity that survives abortion. But their use appears to beg some of the moral questions at issue: "abortus" implies that the entity is mere aborted tissue, whereas "infant" suggests that it is no different from any other baby. Although one or the other of these two views may be correct, it is preferable not to prejudge the issues, and therefore to use the more neutral term "fetus."[1]

Second, it might be asked whether a fetus can have interests, especially interests independent of those of the pregnant woman. Indeed, given the uniquely close connection between pregnant woman and fetus, it may not be appropriate to distinguish between fetal and maternal interests during the pregnancy itself, or at least to regard them as being in conflict. But when pregnancy ends and the fetus is no longer *in utero,* it is entirely reasonable, as later discussion will confirm, to speak of the fetus as having interests independent of those of the woman. In this discussion, it is assumed that the fetus does have interests, after its emergence from the uterus, at least by virtue of its sentience or its potential for sentience.[2]

Third, the issue here primarily concerns relatively developed fetuses of about five months or more gestational age. Both because of their extreme immaturity and because of the destructive nature of most abortion procedures, very few such fetuses actually do survive abortion, at present; very few continue to show a sustained heartbeat, spontaneous respiration, and muscle movement. However, if and when technological advancements permit the time of viability to be moved downward, then moral questions about entitlements of disposition will also arise for younger fetuses that survive abortion.

Viability itself is not the critical moral criterion for determining the disposition of the fetus, since viability is a technologically dependent criterion, having little or nothing to do with the fetus itself, the pregnant woman, or the maternal/fetal relationship. Nor is viability a morally significant point at which abortions should be forbidden, since it makes little sense to permit abortion only up until the very point when the fetus is able to survive outside the uterus.[3]

Fourth, debates about the alleged personhood of fetuses have so far seemed to be of little help in the controversies over abortion. The discussion that follows assumes that the fetus is not yet a person, although it has the capacity to become one, and a being need not be a person in order to have interests. It also assumes that a prochoice position with respect to abortion is morally justified.

Finally, the investigation of these questions has further implications with respect to the potential use of fetuses for research or therapeutic purposes. However, the focus in this paper will be on the question of the preservation of fetuses for the sake of an ongoing life of their own, rather than for use as research material. For reasons of space, I must set aside the many fascinating and significant issues raised by disputes over the disposition of frozen embryos and about fetal tissue transplants, including conception and abortion for the sake of fetal transplant to a relative. Hence, the purview of this chapter is limited.

The Problem

It is possible to distinguish between two different concepts of abortion: abortion as the termination of the pregnancy, and abortion as the killing of the fetus.[4] Although the two are ordinarily empirically linked, they are conceptually and morally distinct. By emphasizing the first concept, a decision to have an abortion could be interpreted primarily as the choice to have the fetus removed from the uterus to surrender it somewhat in the way that children are surrendered in adoption.[5] On this view, the woman who aborts is entitled to have the fetus removed from her body, but not to have it dead. She may want it dead, but whether,

in general, women who abort want this is not clear; it is a matter
for empirical investigation.

> In some cases, abortions are induced because the continua-
> tion of the pregnancy poses a serious threat to the woman's
> life or health. What is intended then is to restore the woman's
> health or save her life by interrupting the pregnancy, sever-
> ing the tie between the woman and her fetus. If the tie could
> be severed without terminating fetal life, this would be the
> preferred outcome. Fetal death may thus be a foreseen but
> unintended consequence of an abortion that is therapeutic
> for the pregnant woman.[6]

Even if the woman does want the death of the fetus, Sissela Bok
and others have argued that this is not a desire that should be
gratified.

> [W]hile a woman does have the right to an abortion in the
> sense of the termination of her pregnancy, she does not have
> the right to the death of the fetus. The termination of early
> pregnancy carries with it, at present, fetal failure to survive.
> But in later pregnancy, where abortion and death of the fetus
> do not necessarily go together, it is a fallacy to believe that
> a right to the first also implies a right to the second.[7]

Defendants of this view of abortion as removal of the fetus must
acknowledge the practical and moral difficulties that attempting
to save aborted fetuses could raise, especially with regard to the
possible survival of injured and damaged fetuses. The difficulties
of "salvaging" very immature fetuses, the subsequent pain caused
to them, and possible resulting disabilities, should not be under-
estimated.

By contrast, other philosophers have emphasized the second
concept of abortion, the killing of the fetus, and have suggested
that the death of the fetus is an inherent component of full abortion rights. "The quest for abortion rights for women is not merely
a quest for control of one's body, though it is surely this in part.

The quest for abortion rights is also a quest for the right to terminate the development of an unwanted foetus[,] which cannot be accomplished without killing it."[8] Hence the questions posed at the beginning of this chapter: Should the pregnant woman be entitled to be assured of the fetus' death, either during or after the abortion? Is it wrong to preserve, against the woman's wishes, fetuses that do or can survive abortion? Since, in general, depriving a being of life (even when that being is clearly not a person) appears to be prima facie wrong, and calls for justification, the burden of proof rests on those who assert an entitlement to the fetus' death. The remainder of this chapter presents and evaluates some major arguments, primarily but not exclusively advanced from a feminist perspective, which suggest that the failure to observe the wishes of the biological mother with respect to the death of the fetus is morally wrong.

The Arguments

Argument 1

To keep the fetus alive against the wishes of its mother is a violation of the woman's reproductive autonomy, in a social rather than a biological sense of that term. A woman who aborts wants not only not to be pregnant but also not to be the mother of this particular fetus, and hence, to the child it could become.[9] For reasons having to do with her economic situation, her health, her psychological state, her job, her education, her relationships, or her commitments to other children, the woman wants to prevent the possible existence of the child that this fetus could become. The entitlement to abort is the entitlement to decide whether or not to be a mother (to the child that this fetus could become).[10]

Respect for the woman's reproductive autonomy, therefore, precludes both "saving" the fetus against the woman's wishes, as well as the (for now) science fiction scenario of transplantation of the fetus to another woman's uterus or to an artificial uterus (so-called "ectogenesis"). As Steven Ross points out, if women going to abortion clinics were told that their fetus would be removed from the uterus, without harm to them or to the fetus, and kept alive elsewhere for the rest of gestation, many would not be satisfied.

> What they want is not to be saved from the "inconvenience of pregnancy" or "the task of raising a certain (existing) child;" what they want is *not to be parents,* that is they do not want there to be a child they fail or succeed in raising. Far from this being "exactly like" abandonment, they abort precisely to avoid being among those who later abandon. They cannot be satisfied *unless* the fetus is killed; nothing else will do.[11]

Moreover, under present-day circumstances, attempts to save the fetus born alive after an abortion "do[es] not prevent a woman from having an abortion but [they] threaten[s] her with a brain-damaged infant".[12] The woman who seeks to end her reproductive activity through abortion is thereby forced to become the mother of an infant that may well be badly damaged by the abortion procedure itself.

Response to Argument 1

In having the abortion, the woman has surrendered the fetus; she has chosen not to become a social mother to it. Hence, if the fetus is saved, she ought not to be and cannot be compelled to become the social mother of the fetus, whatever its medical condition turns out to be. It is no longer her responsibility to rescue, preserve, or nurture it; she does not any longer have any ongoing special moral commitment to it, just by virtue of being its biological mother. If this is acknowledged, then the woman's reproductive autonomy, in particular, whether or not she will become a mother in the social sense, would not be violated by saving the fetus. She would remain only the genetic mother of the fetus, a connection that is unavoidable once conception has occurred.

What is not clear, however, is whether this response unjustifiably trivializes genetic parenthood. The point of Argument 1 is that if the fetus is saved, there will now exist a human being genetically related to the woman.[13] In an important way, this human being is no mere stranger. As Ross argues, this child could represent a kind of failure for the woman, a failure to "be a certain kind of *person,* that is, the sort who has children only when able to raise them oneself in an environment one finds right."[14] Such a woman may feel and believe that she should be the one to raise any children she has borne; instead, if the fetus is preserved, "[t]here would always be in the world a person to whom one was failing to be a proper or full parent, and this is a failure one understandably dreads." In addition, saving the fetus would mean that the biological mother would be subjected to a form of compulsory adoption, with all of the potential for suffering that is experienced by women who give up offspring. "Although we [the pregnant woman] would not be bringing the child up, because someone else (let us assume) is all too gladly embracing those tasks, we do not want precisely this state of affairs to come about."[15]

Argument 2

Saving the fetus against the mother's will is like compelling her to donate organs, blood, or gametes against her will. If the compulsory "donation"—"procurement" is probably a more appropriate word[16]—of bodily parts and products such as organs, blood, or gametes, is neither morally justified nor legally permitted, then fetuses, which are equally body products, ought not to be taken from women against their will.[17] To save the fetus against the mother's will violates a fundamental principle of medical ethics: informed consent. This argument becomes especially urgent if we imagine that fetal viability is pushed back, through technological advancements in neonatal intensive care, earlier and earlier in the process of gestation. Such fetuses could then be "rescued" after abortion, even when there is no evidence for their sentience. Indeed, the case for such a "rescue" has already been advanced by antiabortionists:

Work has already been done toward the development of artificial placentas, and [Bernard] Nathanson sees the possibilities for the rescue of embryos prefigured in the remarkable advent of fiber optics and microsurgery. ...What remains, then, as far as the embryo is concerned, is the development of "an instrument of sufficient delicacy that it can be threaded through the hysteroscope...and can then pluck [the new being] off the wall of the uterus like a helicopter rescuing a stranded mountain climber."[18]

Hadley Arkes predicts that "the law" could compel a woman to have her embryo removed in the first few weeks of pregnancy, once technology reaches the point where it is "possible to rescue the child."[19]

This language of "mountain climbers" and "rescue" suggests, falsely, that a small but sentient and threatened person cowers in the woman's uterus, awaiting its salvation by benign medical technology. It is no coincidence that antiabortionists interpret their own efforts to prevent abortions as the "rescue" of babies.[20] From the point of view taken here, that the fetus is not a person, Arkes' proposal sounds like assault on and invasion of a woman's body, and theft of a component of it. Women would effectively be coerced into being "fetus farms," and women's bodily control would be severely compromised.

Response to Argument 2

Just as people ought not to be compelled to undergo surgery against their will, so also abortions ought never to be compulsory. Women's right to refuse medical interventions on their own bodies must be respected. But providing that the abortion itself is freely chosen, not compelled, it is in this respect not analogous to compulsory organ "donation" or procurement.

Further, there are limits to the organ analogy: Fetuses are in women's bodies, but not part of their bodies . They are arguably not a renewable resource in the same way that blood and sperm are. Unlike bodily organs they can, later in their development, survive independently of the woman's body and become persons, and, at least late in gestation, they are sentient. "The fetus is a developing being and potential member of the human community;" it has "a unique genetic identity, a species-specific physical appearance, and a truncated participation in human social relations."[21] Pregnant women do not seem to experience the fetus as just another bodily organ; they often experience the fetus as both part of and different from themselves, and they sometimes develop a type of relationship with the fetus before its birth. Yet human beings do not usually develop relationships with their bodily organs. In addition, there is a disanalogy insofar as while organs serve some purpose in the person's body, so that if they are removed for nontherapeutic reasons there is a deficit in the body, this is not the case for the fetus: the woman needs her bodily organs, but she does not need the fetus; the fetus needs her. These characteristics of fetuses, as opposed to bodily organs, provide all the more reason to be concerned about the fate of the fetus that survives abortion, and they discredit the analogy between fetuses and bodily organs.

In addition, the analogy to compulsory organ procurement appears, unjustifiably, to imply that the woman owns the fetus: "In this paradoxical morality there [is] a curious assertion of 'property rights:' it was somehow easier to kill the fetus in the womb than to give away to others what was recognizably a child— and recognizably, also, a child of one's 'own.'"[22]

Interestingly, some commentators have been willing to see fetuses, along with organs, limbs, and bodily fluids, as property,[23] property over which some limitations of disposal may apply. But even if organs, limbs, and fluids should be seen as property (and there may be good arguments against it), the disanalogies between fetuses and bodily parts suggest that fetuses cannot likewise be regarded as property.

The purview of informed consent is justifiably limited with respect to blood and gametes, and perhaps it ought likewise to be limited with respect to fetuses. We are not entitled to say who ought or ought not to receive our blood or gametes. We are not allowed to make invidious exclusions with respect to their use; for example, racists are not entitled to specify that only whites may have their blood; the homophobic are not entitled to specify that only heterosexuals may have their gametes (although this latter principle often governs the practices of sperm banks). A comparable limitation on the range of informed consent with respect to the disposition of fetuses may, therefore, seem morally justified.

Nevertheless, there is a significant moral and social policy question concerning whether we should, in fact, have more control over the disposition of bodily parts and products. In a recent, much disputed case, a man's cancerous blood and tissue were surgically removed and subsequently cultured and developed, allegedly without his knowledge or explicit consent, to develop a patented and commercially valuable cell line.[24] The judicial assessment of the case manifested "the irony of the conclusion that everyone *except* the patient can own the patient's removed cells and treat them as property."[25] Such disputes at least indicate that it may be important to grant to individuals more knowledge and decision-making about and control over the disposition of materials removed from their bodies, even if that material merely seems to the patient to be "waste."

The case against the saving of fetuses that survive abortion need not rest on seeing the fetus as property in the same way that jewelry or other possessions are property. To say that "x is mine" is sometimes to make an ownership claim, but it is sometimes, instead, to claim x as my responsibility, or as subject to my decision-making. For example, when a woman describes offspring as "my children," she is not claiming that she owns them; rather she is asserting responsibility for and connection to them. Thus, Susan Sherwin argues that

>...women are in a privileged position with respect to the fetuses developing in their bodies, and...in most circumstances, they are entitled to decide the future of those fetuses. This is not

because they own the fetuses, for they ought not to be free to sell them, but because they are responsible for them and should be trusted to decide if continued life when removed from the womb is in the best interest of the fetus.[26]

Thus, rejecting the compulsory preservation of fetuses need not assume a property relationship between the pregnant woman and the fetus. When a woman chooses abortion, she does not necessarily choose to have her fetus "snatched" from her; if she does not, then preserving it is in that respect comparable to a compulsory organ "donation" in which the patient chooses organ removal but does not agree to the subsequent salvaging and use of the organ.[27]

Argument 3

By virtue of her physical relationship to it, the biological mother is the best person—perhaps the only person—suited to deciding the disposition of the fetus. Mary Anne Coffey states, "If her child is dying of a fatal illness...a mother now can direct that the child not be resuscitated. Why deny a woman the same right when a fetus that survives an abortion dies?"[28] Sherwin adds that it is the woman's "vision of threats facing the developing child" that might motivate her to want the fetus destroyed:

> An analysis attentive to the interests of children and women...must recognize that protecting the interests of the embryo does not necessarily mean preserving its life.... [I]t is legitimate...that the person who has the most intimate relationship with the fetus and who has the most invested in its development—i.e., the mother—should be the one to decide on how its interests may best be served.[29]

Response to Argument 3

It must be acknowledged that the pregnant woman is the best and only person to make decisions about the fetus while it is *in utero*. But to the extent that a fetus *ex utero* is comparable to a premature baby, as argument 3 appears to grant, the interests of the offspring should be the prevailing criterion for decision-making about it. It is improbable to suppose that a fetus that survives abortion is always better off dead (though it may often be[30]); moreover, there may be adoptive parents willing to raise it. Hence, giving the biological mother the entitlement to the death of the fetus would sometimes mean overlooking the fetus' interests in a way that would not be condoned for premature infants; such a choice would not be morally justified.

Should the fetus that survives abortion be regarded as an ordinary newborn, or rather, as a premature newborn? In terms of its physical characteristics it may be very like such a newborn, providing it has not suffered injuries in the process of the abortion. Its "arrival" into the world is induced rather than the outcome of the natural course of labor—but so are many other births, which are induced through the use of pitocin drips or caesarean sections. Moreover, like an ordinary newborn, the fetus that survives abortion "may have no intrinsic properties that can ground a moral right to life stronger than that of a fetus just before birth, [but] its emergence into the social world makes it appropriate to treat it as if it had such a stronger right."[31] The point here is *not* that the fact of birth (or removal from the uterus during abortion) constitutes the fetus as a person, but

> ...it does become a biologically separate human being. As such, it can be known and cared for as a particular individual. It can also be vigorously protected without negating the basic rights of women. There are circumstances in which infanticide may be the best of a bad set of options. But our own society has both the ability and the desire to protect infants, and there is no reason why we should not do so.[32]

Argument 4

Deliberately withholding the determination of the disposition of the fetus from the biological mother is yet another example of the takeover of reproduction from women. "The extent to which the rights of women are diminished in abortion policy and litigation, when the fetus is part of the woman's body, should make us seriously question the extent to which they will be further diminished as the fetus is removed from the female body."[33] Sherwin argues that without the freedom to decide the fate of their fetuses, "women will not have the reproductive freedom necessary, and, in particular, they will certainly have difficulty in getting abortions."[34] Anne Donchin argues that the technological maintenance, outside the woman's body, of fetuses that survive abortion would be a manifestation of distrust of women's bodies. Laboratory technicians are not likely to do "a more competent job of gestation than pregnant women."

> And if extrauterine gestation were to become an established practice, would not many women be pressured to adopt it— "for the good of their baby?"

> Though abortion may count as a harm to the fetus, laboratory gestation would as well—not only to particular "unwanted" fetuses but to all future fetuses. For, within the prevailing social framework, once the practice was established it is unlikely that only intentionally aborted fetuses would be nourished in laboratories. Any other fetus considered "at risk" for any reason would count as a potential beneficiary of laboratory observation and intervention.[35]

Response to Argument 4

It must be granted that the appropriation of reproductive control from women must be resisted. To this end, women are entitled to have the type of abortion they choose (within the limits of good medical practice). This means that they are entitled to choose, if they wish, means of abortion that will likely produce the death of the fetus; they are not required to choose abortifacients that will preserve the life of the fetus. Pregnant women are not morally required to exhibit "moral heroism" by putting their own lives at risk for the sake of a possibly viable fetus.[36] Nor are women compelled to undergo less safe forms of abortion in order to provide intact fetal tissue for purposes of transplant or research. The availability of these choices and protections with regard to the process or means of abortion respects women's bodily autonomy. In addition, the encroachment of the state and the medical profession on women's reproductive autonomy during pregnancy must be adamantly resisted,[37] along with notorious legal attempts to compel prenatal treatment, administer forced caesareans, or "take custody of the fetus" before birth.[38]

It is, therefore, necessary to reject the views of philosophers such as David S. Levin, who claims that if there is the possibility of keeping a fetus alive after it is removed from a woman's body, then she has a "minimal responsibility" of allowing the being to be removed alive.[39] Levin states, "if and when removal without killing becomes possible, [the pregnant woman's] right to control her own body cannot justify killing the fetus."[40] The phrase "killing the fetus," though, is ambiguous without any specification of time or place. If it means killing the fetus *in utero,* as a consequence of a particular abortion operation, then the killing is justifiable by reference to the woman's control over her own body. Only if the killing of the fetus takes place after it has been removed from the woman's body is it no longer justified just by reference to the woman's control over her own body, or to the fact that she is the biological mother of the fetus.

Conclusion

The examination of the alleged rights of biological mothers to determine the disposition of the fetus after abortion suggests that it is essential to make a distinction between two different questions: First, who should decide about the disposition of the fetus? Second, does the pregnant woman who seeks an abortion have an entitlement to the death of the fetus?

As the discussion has suggested, the pregnant woman should decide the disposition of the fetus. For the pregnant woman is, in Sherwin's words, "the person who has the most intimate relationship with the fetus and who has the most invested in its development."[41] Given the history of the appropriation of women's reproductive autonomy by male partners and by members of the medical establishment, it is deeply problematic to assign this responsibility to the biological father or to physicians. As John Robertson suggests, "[I]n cases of conflict between [the pregnant woman] and the father over disposition, one could argue that her interests control because the fetus was removed from her body."[42]

But whereas no other person than the mother has a greater entitlement to decide about the fate of the fetus that survives abortion, and the decision therefore belongs to her, it does not follow that the woman's decision, by virtue of its being hers, is necessarily correct. In particular, the choice of death for the fetus is not rendered morally correct simply because the decision is made by the biological mother of the fetus. Though the pregnant woman is entitled to forms of abortion that may result in the death of the fetus *in utero,* she does not have an entitlement to the death of the fetus if it survives abortion. The arguments canvassed in this chapter do not establish the truth of the claim that the woman is automatically entitled by virtue of being the biological mother to have the fetus die after it is removed from her uterus, or that her wishes about its disposition, where that disposition is not in the fetus' interests, are necessarily morally justified.

What then should be done with respect to fetuses that survive abortion? The answer is that their mothers' decisions on their behalf should be guided by the interests of the fetuses, just as they would for other premature infants. This is not to say that fetuses must inevitably be preserved and protected, or that extraordinary medical measures must necessarily be taken on their behalf, for sometimes the fetus that survives abortion is better off dead. Nor is it to say that women should be compelled to raise the fetuses they have aborted, for a decision to abort is (in part) a considered choice not to be the mother of the child this fetus could become. Nor, finally, is it to say that the decision about the disposition of fetuses must devolve on the male progenitor or the physician(s), for there is no case for their entitlement that overrides that of the pregnant woman.

It is to say, though, that the burden of proof still rests on those individuals, including biological mothers, who wish to kill the fetus, to let it die, or to treat it in ways that are not in its interests.

Notes and References

[1]Moreover, there is a growing body of bioethical literature that uses the term "fetus" to refer to the entity that survives abortion.

[2]On the significance of sentience, *see* M. A. Warren (1989) The moral significance of birth. *Hypatia* **4(3)**, 49–52.

[3]C. Overall (1987) Chapter 4, in *Ethics and Human Reproduction: A Feminist Analysis.* Allen and Unwin, Boston, MA.

[4]C. Overall (1987) *Ethics and Human Reproduction: A Feminist Analysis.* Allen and Unwin, Boston, MA.

[5]R. M. Herbenick (1975) Remarks on abortion, abandonment, and adoption opportunities. *Philosophy and Public Affairs* **5(1)**, 98–104. This analogy may not be completely appropriate in cases of abortion for fetal abnormality. In such cases, the fetus is very much wanted, yet the fetus itself may be better off dead, not preserved. On the ambiguities of seeking abortion for the benefit of the fetus, *see* P. F. Camenisch (1983) Abortion: For the fetus's own sake? in *Medical Ethics and Human Life* (J. E. Thomas, ed.) Samuel Stevens, Toronto, pp. 135–143.

[6]M. B. Mahowald, R. A. Ratcheson, and J. Silver. (1987) The ethical options in transplanting fetal tissue. *Hastings Center Report* **17 (1)**, 13.

[7]S. Bok (1984) The unwanted child: Caring for the fetus born alive after an abortion, in *Cases in Bioethics,* revised ed. (C. Levine and R. M. Veatch, eds.), Hastings Center, Hastings-on-Hudson, NY, p. 2.

[8]D. S. Levin (1985) Thomson and the current state of the abortion controversy. *Journal of Applied Ethics* **2(1)**, 125; cf. S. L. Ross (1982)

Abortion and the death of the fetus. *Philosophy and Public Affairs* **11(3)**, 236; R. M. Herbenick (1975) Remarks on abortion, abandonment, and adoption opportunities. *Philosophy and Public Affairs* **5(1)**, 101.

[9]I owe this argument to Lois Pineau.

[10]D. I. Wikler (1979) Ought we to try to save aborted fetuses? *Ethics* **90**, 58–65.

[11]S. L. Ross (1982) Abortion and the death of the fetus. *Philosophy and Public Affairs* **11(3)**, 238, his emphasis.

[12]D. C. Nathan (1984) The unwanted child: Caring for the fetus born alive after an abortion, in *Cases in Bioethics,* revised ed. (C. Levine and R. M. Veatch, eds.), Hastings Center, Hastings-on-Hudson, NY, p. 4.

[13]For further discussion of the interest in avoiding genetic offspring, *see* J. A. Robertson (1989) Resolving disputes over frozen embryos. *Hastings Center Report* **19(6)**, 7–12.

[14]S. L. Ross (1982) Abortion and the death of the fetus. *Philosophy and Public Affairs* **11(3)**, 241, his emphasis.

[15]S. L. Ross (1982) Abortion and the death of the fetus. *Philosophy and Public Affairs* **11(3)**, 239.

[16]J. G. Raymond (1990) Reproductive gifts and gift giving: The altruistic woman. *Hastings Center Report* **20(6)**, 7–11.

[17]A version of this argument was presented to me by Sanda Rodgers.

[18]H. Arkes (1986) *First Things: An Inquiry into the First Principles of Morals and Justice.* Princeton University Press, Princeton, NJ, p. 377.

[19]H. Arkes (1986) *First Things: An Inquiry into the First Principles of Morals and Justice.* Princeton University Press, Princeton, NJ, p. 378.

[20]G. Leber (1989) We must rescue them. *Hastings Center Report* **19(6)**, 26–27.

[21]K. Nolan (1988) Genug ist genug: A fetus is not a kidney. *Hastings Center Report* **18(6)**, 16.

[22]H. Arkes (1986) *First Things: An Inquiry into the First Principles of Morals and Justice.* Princeton University Press, Princeton, NJ, p. 371.

[23]L. B. Andrews (1986) My body, my property. *Hastings Center Report* **16(5)**, 28–38.

[24]L. B. Andrews (1986) My body, my property. *Hastings Center Report* **16(5)**, 28–38; G. J. Annas (1988) Whose waste is it anyway? The case of John Moore. *Hastings Center Report* **18(5)**, 37–39; G. J. Annas (1990) Outrageous fortune: Selling other people's cells. *Hastings Center Report* **20(6)**, 36–39.

[25]G. J. Annas (1990) Outrageous fortune: Selling other people's cells. *Hastings Center Report* **20(6)**, 37, his emphasis.

[26]S. Sherwin (1988) Review of *Ethics and Human Reproduction: A Feminist Analysis. Atlantis* **13(2)**, 125.

[27]There are further feminist reasons for avoiding the ownership paradigm for the fetus, since there is a developing history of seeing the fetus as the property of the male progenitor, the man's "baby."

[28]M. A. Coffey (1989) Review of *Ethics and Human Reproduction: A Feminist Analysis. Resources For Feminist Research/Documentation sur la recherche feministe* **18(1)**, 11.

[29]S. Sherwin (1988) Review of *Ethics and Human Reproduction: A Feminist Analysis. Atlantis* **13(2)**, 125.

[30]S. Sherwin (1988) Review of *Ethics and Human Reproduction: A Feminist Analysis. Atlantis* **13(2)**, 123–125.

[31]M. A. Warren (1989) The moral significance of birth. *Hypatia* **4(3)**, 57.

[32]M. A. Warren (1989) The moral significance of birth. *Hypatia* **4(3)**, 62.

[33]J. G. Raymond (1990) Of ice and men: The big chill over women's reproductive rights. *Issues in Reproductive and Genetic Engineering: Journal of International Feminist Analysis* **3(1)**, 49.

[34]S. Sherwin (1988) Review of *Ethics and Human Reproduction: A Feminist Analysis. Atlantis* **13(2)**, 125.

[35]A. Donchin (1989) The growing feminist debate over the new reproductive technologies. *Hypatia* **4(3)**, 144.

[36]L. Walters (1984) The unwanted child: Caring for the fetus born alive after an abortion, in *Cases in Bioethics,* revised ed. (C. Levine and R. M. Veatch, eds.), Hastings Center, Hastings-on-Hudson, NY p. 6.

[37]J. Gallagher (1989) Fetus as patient, in *Reproductive Laws for the 1990s* (S. Cohen and N. Taub, eds.), Humana Press, Clifton, NJ, pp. 185–235; National Association of Women and the Law (1989) A response to *Crimes Against the Foetus,* The Law Reform Commission of Canada's working paper #58. Ottawa, Ontario.

[38]M. Thompson (1988) Whose womb is it anyway? *Healthsharing* **(Spring)**, 14–17.

[39]D. S. Levin (1985) Thomson and the current state of the abortion controversy. *Journal of Applied Ethics* **2(1)**, 124.

[40]D. S. Levin (1985) Thomson and the current state of the abortion controversy. *Journal of Applied Ethics* **2(1)**, 125; cf. E. F. Paul and J. Paul. (1979) Self-ownership, abortion and infanticide. *Journal of Medical Ethics* **5**, 135.

[41]S. Sherwin (1988) Review of *Ethics and Human Reproduction: A Feminist Analysis. Atlantis* **13(2)**, 125.

[42]J. A. Robertson (1988) Rights, symbolism, and public policy in fetal tissue transplants. *Hastings Center Report* **18(6)**, 9.

Frozen Embryos
and Frozen Concepts

Wade L. Robison

The case of *Davis* v. *Davis* made headlines around the world.
The facts were simple. Mr. and Mrs. Davis had been unable to
conceive, and Mrs. Davis had undertaken "in vitro procedures,"
as the court put it, to attempt to produce a child.[1] Eggs were taken
from her and inseminated with sperm from Mr. Davis, and, of the
resulting fertilized eggs, two were implanted immediately without effect and seven frozen for later implantation. The couple,
meanwhile, decided to divorce, and the issue between them of
importance for the court concerned the disposition of the seven
frozen embryos.

The court decided that temporary custody should be given to
Mrs. Davis "for the purpose of implantation,"[2] and that the reason
she should be given custody was, first, that she and her husband
had succeeded in producing "children, in vitro," and second, that
"it is to the manifest best interest of the children, in vitro, that they
be made available for implantation to assure their opportunity for
a live birth."[3]

The case raises a variety of issues, from the question of what
rights a man has over eggs his sperm have made fertile to the
question of what would happen should Mrs. Davis decide not to

From: *Biomedical Ethics Reviews • 1991*
Eds.: J. Humber & R. Almeder ©1991 The Humana Press Inc., Totowa, NJ

implant the eggs. The issue that needs settling before any other issue can be properly settled, though, concerns the status of the frozen embryos: What are they?

The court's decision that they are "children, in vitro" and, thus, that Mr. and Mrs. Davis "have accomplished their original intent to produce a human being to be known as their child" should strike a casual reader as bizarre.[4] When a court declares that an abused child has not been proven abused beyond a reasonable doubt and, therefore, the child is not abused, in law, we understand the court to say it cannot find the child abused, even if the child is. We distinguish readily between a statement of law and a statement of fact, and we may not mind the fiction, though we may mind the judgment. But when the court in *Davis* v. *Davis* declared that the seven frozen embryos are "children, [even if] in vitro," the line between statements of law and fact is strained. As we shall see, the form of the court's argument makes it difficult to know how else to read its conclusion but as a statement of fact. Those frozen embryos are children, the court is saying, and that is not only false, but so wildly at variance with the facts as to be bizarre, ready material for jokes.

The question "What are they?" concerns the status of these entities, and has two aspects. We need to know what these beings are ontologically—as beings in the world, as human beings, dogs, and butterflies are beings in the world—and we need to know what these beings are morally. What, that is, is their moral status? How are we to respond to them, if at all, and respond to their interests, if they are beings capable of having interests and have them? The court concludes they have interests in being born, and thus, interests in being implanted, and we shall need to determine whether they are the kinds of beings who can have such interests. The concern with their ontological status is a concern with whatever natural features they may have that bear on their moral status, if any.

The Court's Arguments

The court begins by asking what it calls "the most poignant question of the case: When does human life begin?"[5] It then turns to four questions it thinks must first be answered: "Are the embryos human? Does a difference exist between a preembryo and an embryo? Are the embryos beings? Are the embryos property that may become human beings?"[6]

Some of these questions seem odd, and the string is odd. Why is the court asking whether the embryos are beings? What could that mean? The court's aims are clarified, though, by the way it cuts through these questions with a single line of argument. The DNA-coding of each frozen embryo makes it human, it claims, and makes it unique. The court, thus, concludes that "the life codes for each special, unique individual are resident at conception and antimate [sic] the new person very soon after fertilization occurs,"[7] and thus, that the Davises have produced "a human being to be known as their child."[8] They have produced seven.

This is a bit quick, but we need to understand that the court had a dilemma. Those getting a divorce have to dispose of property, which is to be split between them, and may have children whose custody must be taken care of. There is nothing else between a married couple except children and property that is a matter for the court's concern. So the court's questions presuppose that distinction: The court may take judicial notice of property and of children, and nothing else. As the court notes, "The court must find the facts and apply the law in order to make an equitable distribution of the parties' property, if any, and provide for the care, custody and support of the parties' minor children, if any."[9] So, the embryos are going to be property or children, and nothing else, if they can properly be disposed of by the court, and it is in light of this choice that the court asks whether the embryos are beings or "property that may become human beings."

It is also in light of that constraint, I think, that the court rejects a distinction some of the expert witnesses insisted on between an embryo and a preembryo, which the court defines as "the human entity existing before the passage of fourteen days of development, prior to attachment to the uterine wall and the development of the primitive streak."[10] The distinction allows a different status for the frozen embryos than that accorded embryos. But, the court says, "the term 'preembryo' serves as a false distinguishing term in this case."[11] The court has no third category in which to put such entities: They are either children or property. Calling them preembryos to distinguish them from embryos makes assimilating them to the category of children more difficult and considering them as property easier. But, the court argues, the attempted distinction is misplaced: Since the DNA-coding exists from conception, and since that is crucial to determining that something is a being and what kind of being it is, any distinction between stages of that being makes no real difference to its status. It is a human being, the court claims, because of its DNA-coding.

This claim is intended as empirical. The court appeals to the evidence of scientists, and it presents its conclusion, and its objection to the alleged distinction between embryos and preembryos, as founded on scientific evidence, "the technical arguments of human genetics."[12] But the court goes from that claim to claim further that any such human being is a "new person very soon after"[13] and that not only are these beings not property, each is a being that Mr. and Mrs. Davis intended to produce, namely, "a human being to be known as their child."[14]

What powers these further claims is that the court must treat the embryos as children or property. It is that additional premise, and the court's unwillingness to treat as property what it has just determined to be human beings, that drives it to call the frozen embryos children. The legal finding thus comes off as a bizarre empirical claim.

Yet some of the court's claims are hedged in their original introduction. The court does not say that the frozen embryos are persons, but that they will be persons "very soon after," the "soon after" implying that the embryos are not yet persons, the "very" making it unclear what more may be needed to make them persons. The phrase "to be known as their child" does not imply that the being is their child. It says that the entity will be known as their child—perhaps meaning only that if it is implanted, and takes, and goes to term, and is born live, then it will become their child. Later in its argument, the court refers to the frozen embryos not as children, but as "children, in vitro," yet it awards custody to Mrs. Davis presumably because the frozen embryos are children. At one point, the court refers to "the practical storage life" of the embryos, as one might refer to the shelf-life of tomatoes, and yet it concludes by saying that it will reserve the issue of final custody until such time "as one or more of the seven cryogenetically preserved human embryos are the product of live birth," thus, assuming the essential identity of the entity from frozen state to infant.[15]

So, the court's language and decision reflect unsureness about the status of these beings. It is unwilling to treat them as property, must treat them as property or children, and yet is not quite ready to call them children *simpliciter,* without a "storage life." The phrase "children, in vitro" rather nicely captures the court's ambivalence—the phrase not modifying the kind of being they are, but telling us that they are somehow in a special position, making it unclear how we are to respond to them. They are children, but in this somewhat awkward ontological and moral place.

Puzzling Implications

We can begin to fix an entity's moral status, in a way Leibniz would approve, by determining its moral relations to other entities. Fix an entity's moral relations, and one will have a complete understanding of the entity's status in a moral hierarchy. One will know how it relates, and thus, compares to all other moral entities how, for instance, it fares when its interests conflict with any other moral entity's.

The court's decision has implications about how to settle some legal disputes and their moral import. We can thus begin to get a fix on what the court takes to be the moral status of those frozen embryos by seeing what the court says about their legal status. The status puzzles as much as the implications.

First, the court gave Mrs. Davis "temporary custody...for the purpose of implantation,"[21] claiming that it "serves the best interests of the child or children, in vitro, for their Mother, Mrs. Davis, to be permitted the opportunity to bring them to term through implantation."[22] She is not obligated to have any implanted, though, and so the question arises, "What happens if she should decide not to avail herself of this opportunity?"

That issue was not before the court, but the decision makes it difficult to see what the answer could be. If the frozen embryos are already children, even if in vitro, it is arguable they have a right to be implanted. If it serves their "best interests" to be implanted because, as the court notes, "no one disputes...that unless...[they] are implanted, their lives will be lost,"[23] then they arguably have more than a passing interest, and more than an interest, in being implanted. If they are not implanted, "they will die a passive death,"[24] the court says, and no one presumably denies that a child, who happens not to be in vitro, would have a right not to die a passive death. Whether this right would outweigh the right of someone else to refuse to aid such a child is a different question, not settled merely by noting that the child would have a right. It is much easier to dismiss a person's interests when they conflict with someone else's rights than to dismiss someone's right to life.

Since the frozen embryos could be implanted in some other womb, they are not condemned to death if Mrs. Davis should decide not to have them implanted. Yet the court does not say that such "children, in vitro," have to be implanted. The choice is between Mrs. Davis implanting them or, short of another court proceeding to give another woman the same opportunity, their dying a passive death. The latter is acceptable, apparently, should Mrs. Davis decide not to implant them and no one else initiates a court action to have the frozen embryos implanted. They, in short, carry no claims with them against anyone—even to further their lives.

This feature of the frozen embryos is consistent with a second puzzle this case presents: if any frozen embryo is implanted, and takes, it may be aborted. Suppose Mrs. Davis decides to have one or two implanted and they develop into fetuses. She is permitted to have an abortion, in accordance with the conditions laid down by Tennessee law. Nothing in that law prohibits someone who became pregnant by having frozen embryos implanted from having an abortion. She is not obligated to carry them to term.

So, when the court calls these frozen embryos "children, in vitro," that categorization does minimal moral work. Presumably, it would be morally wrong to abort children, if fetuses were children, for any but very weighty moral reasons, such as a great risk to the mother's health, just as it would be morally wrong to kill infants for any but very weighty moral reasons. So, because the frozen embryos need not be implanted, and once they are implanted, need not be taken to term, but may be aborted, the court, in calling the frozen embryos "children, in vitro," is not putting them into a category that gives the embryos the rights and privileges of the children we all know and love.

Calling them "children, in vitro," prevents them from being categorized as property. So, neither Mrs. nor Mr. Davis are permitted to dispose of them in the various ways in which they may dispose of their joint property—their lawn mower, their house, the money in their savings account. But, since calling them "children, in vitro," does not even guarantee that they will be implanted, that categorization serves little more moral work than to prevent their being sold at a yard sale.

The court's language is, thus, misleading, to say the least. The frozen embryos have the status of children, because otherwise they would have the status of property. But, when one examines what the court says and implies about their relations with those about them, they have no status comparable to that of children. Still, by saying that the frozen embryos have an interest in being implanted, the court implies that they have *some* moral status, and we need to turn to consider that and to consider how one determines the moral status of any entity.

Independence of Considerations

The court spent a great part of its labors on the claimed distinction between embryos and preembryos, and it "is aware," it says, "that many members of the public consider the questions involved to be of a strict moral nature, not of a legal nature, and that the impact of the court's decision may offend, give support to or otherwise affect many moral views of a substantial segment of the public."[20]

Categorizing something may have moral implications. If my dog is "a member of the family," its moral status from a guard dog's. One may deeply regret leaving the former out all night, and feel shame at such neglect, whereas not being at all concerned, except for the possible loss of protection, from leaving out one's guard dog.

So, we should not suppose that categorizing the frozen embryos is a morally neutral matter: It may not be. But, what we ought to suppose is that any decision about their ontological status be made as cleanly as possible on nonmoral grounds. Moral considerations should not drive that categorization. Even if many may be offended if the frozen embryos are not categorized as humans, we ought to be concerned to get exact the ontological status of frozen embryos—what are they among all the things that exist in the world?—without paying attention to the moral implications of any categorization.

I say this, but even what may appear to be the most objective of investigations may well be loaded with subjective baggage. I do not want to say something like this: Deep moral commitments are shared, and we can, therefore, act within them, making distinctions between more shallow commitments to value and what we call scientifically objective judgments. To say that would presuppose that those deeper commitments do not permeate and change our immediate contact with the world.

I rather want to say that we should take it as an heuristic ideal, a regulative principle, to attempt to make such decisions about, e.g., the status of frozen embryos independently of any value commitments. That principle will make some attempts inappropriate, though it will not guarantee that the categorization is value-neutral. Nothing can guarantee that, but it will help us make sure that we make explicit whatever value judgments we are making.

I think that the court was pursuing this principle when it decided that the "term 'preembryo' serves as a false distinguishing term in this case."[21] It quoted the *Report of the Ethics Committee of The American Fertility Society,* which defined a preembryo as "a product of gametic union from fertilization to the appearance of the embryonic axis."[22] "The preembryonic stage," the report continues, "is considered to last until 14 days after fertilization."

The court might have asked, "Stage of what?" The preembryonic stage is marked by DNA-coding and by "its potential to become a person,"[23] and those features mark the embryo as well: The embryo is just farther along. So, the distinction is no more between kinds of beings, the court claims, than the distinction between an old person and a young one. The being is *of the same kind,* but in different stages of development.

So, the court objects to the attempted distinction between preembryo and embryo because, even though the committee report, as quoted by the court, said that the distinction was "not intended to imply a moral evaluation of the preembyo,"[24] the court thinks the distinction of value only to allow a different categorization of the frozen embryos and, thus, a different status—something not a child, even if not property. It is not germane to what is at issue: What are these frozen embryos (ontologically)? The court's answer, however, is a puzzle. It makes six different claims about the frozen embryos, some explicitly and some implicitly:

1. They are DNA-coded human.[25]
2. They are human embryos.[26]
3. They are human beings.[27]
4. They have the potential to be persons.[28]
5. They are "very soon" to be persons.[29]
6. They are "children, in vitro."[30]

The court uses these terms almost interchangeably. In the paragraph after considering at great length the disposition of "the children, in vitro," it goes on to vest "temporary custody of the parties' seven cryogenically preserved human embryos...in Mrs. Davis."[31] Yet these are very different ways of describing the embryos, and the court has failed to explain how it proceeds from marking them out as a distinct kind of entity—DNA-coded human—to inferring that they are human beings, or each a "new person very soon after fertilization occurs,"[32] or children, even if in vitro. The court has failed to set out the moral premises that justify its move from what is ostensibly, at least, a scientific endeavor to the moves that so classify these entities that, by the mode of classification, they have some moral status.

If having a particular DNA-coding makes an entity human, as the court suggests, then that need have no moral implications *in and of itself:* One needs to add some premise or premises about the value of such beings. The court gives none, but, sliding from one category to the next, presupposes they have moral status of some sort. It makes a difference to how we *think* of these entities if they are children or if they are DNA-coded human, capable of becoming children. The court's categorizations of these entities is, thus, not only sloppy, but driven by unstated moral assumptions. I will assume that the frozen embryos are DNA-coded humans. Treated as a claim about their biology, such an assumption makes no moral presumptions, I would argue, and one can ask, from that baseline, what moral additions are necessary to justify categorizing these entities in any other way—as human beings, for instance, or as "children, in vitro." I shall take as my stalking horse for the rest of this chapter, the claim that the frozen embryos are "children, in vitro." That allows me to make most dramatically the points that can be made about any of the other ways the court characterizes the frozen embryos.

Moral Standing

Calling the frozen embryos "children, in vitro" is not a morally innocent move. Making such a judgment, I shall argue, presupposes a moral theory in which such entities have moral standing, and, as we shall see, frozen embryos have moral standing in a theory significantly different from Kant's, for instance, or Mill's.

Consider the following case. In a "Twilight Zone" story, an alien comes to earth, bearing inducements for earthlings to give up war and become fat and prosperous. In the final scene, a newspaper reporter has entered the final gate of a space ship to go to the home planet to see what is cooking with all the people who have left and ceased to write home. His girlfriend, running to catch him, shouts to him that she has translated the book the alien left at the United Nations. "It's a cookbook!"

We are being fattened up to be eaten. Surely, we must think, our being able to communicate with these aliens gives us some claim on them. How can they eat someone they can talk to? We consider ourselves moral entities, with rights not to be eaten, and to consider ourselves as moral entities is to make a commitment to a kind of moral world in which such beings ought to recognize us. We are saying, "We share the same moral universe as you: It is wrong to eat us!" Since we can talk with them, we think they ought to listen to us—and recognize us as kindred moral beings. We think that we have moral standing in their moral world, so that they should take our moral position into account, and we think that if they did, they would not eat us.

When the judge in *Davis* v. *Davis* calls the frozen embryos "children, in vitro," and says that a frozen embryo will become a "new person shortly after fertilization," he is making the same sort of move we might make in such a "Twilight Zone" story. He is trying to include those frozen embryos in our moral universe, the one we share whatever its justification, presuppositions, or implications. If they are children, we cannot readily justify disposing of them any more than aliens could readily eat us if we are the same sorts of beings as they. Like us, the frozen embryos would have standing to object or, rather, to have someone object for them.

To determine *standing* is to determine the crucial claim about the scope of a moral theory, for it tells us who within that theory is entitled, morally, to a hearing. Many entities may bear moral relations—as objects of moral relations, for instance—without being entitled to be heard within a theory. A theory gives moral value to what falls within its scope. For instance, Nozick's theory of justice gives moral value to certain acquisitive characteristics: Their development and effective use makes a moral difference in the world created by his moral theory. We need to distinguish, though, what has moral value because of a theory, and so, is subject to moral relations, and what has standing within a theory. To have standing is fundamentally to be entitled to a hearing. Parsing that out means that a being with standing is, among other things, entitled to be the focus of relations countenanced by the theory—to be entitled to initiate moral relations by promising, for instance, to bear moral relations to others by being the beneficiary of gifts, for example, and to hold the status appropriate for moral agents, subject to responsibility, shame, and guilt those outside the system cannot have.

To have standing is not to imply that one's claims will be recognized. That will depend on the resolution of the particular configuration of relations—what is in conflict and which party has the better claim. Indeed, to have standing is not even to imply that one has rights. All sentient beings have standing in Mill's theory, since the scope of his utility principle covers "the whole sentient creation," but a being with a capacity for pleasure and pain may have no rights at all, only interests.[33] Having standing in Mill's view means counting for one equally with all others who have standing "so far as the nature of things admits."[34]

One can get a sense of the importance of standing by considering *Dred Scott* v. *Sandford*.[35] Scott was a slave taken by his master first to the free state of Illinois and then to the Louisiana Territory, free by the Missouri Compromise. Sometime after his return to Missouri, he sued his master, Sandford, for his freedom. After being denied his freedom by the Missouri Supreme Court, he sued in the Federal District Court, and since his master then lived in Massachusetts, he sued under that Article of the Constitution that gives the Supreme Court original jurisdiction between citizens of different states. When the case was appealed to the Supreme Court, the court puzzled over whether Scott had standing, whether, that is, he could bring suit. If he was a citizen, he could, but he was suing Sandford for his freedom, and if he lost his suit, he would not be a citizen and so would be unable to bring suit. The court was in a conceptual dither: Would hearing the case itself give Scott standing and, thus, settle that he was a citizen?

The court determined that it had before it the question whether the lower court had properly exercised its jurisdiction, and to settle that issue it needed to decide whether Scott was a citizen.

It decided that he was not and that no black who had been a slave or who had been free at the time of the adoption of the Constitution could ever be a citizen. Any rights or privileges any state might have granted a native black[36] were gifts, Chief Justice Taney argued, and those who were granted those rights and privileges were not entitled to them: Any state could take them back at any time.

The effect of *Dred Scott* was that no native blacks could go to federal court to claim any right whatsoever. Not being citizens, and so not having standing to bring suit, they had neither any of the rights, privileges, nor immunities of citizens, nor the capacity to protect such rights, privileges, and immunities if they had had them. They were legal outcasts, subject to the Constitutional system, but unable to use that system—except by some states giving them limited use. They, thus, had a status within the system: Like property, they figured in legal relations. They had no standing, though, within the system: The system legally empowers citizens, and they were not citizens. Put in a stronger, but perhaps misleading way, they might come to have rights and privileges within the system, for individual states might make them the foci of legal relations, but since that status depended on states, not on the Constitution, and they had no national citizenship, they had no right to any rights and privileges they might enjoy, and so, no standing to object if those rights and privileges were denied or abridged.

Similarly, any being cut out of a moral theory may well be subject to that theory, but cannot appeal morally according to that theory. Those empowered by the moral theory are not required by the theory that empowers them to grant the status they are accorded by the theory to any entities cut out of the theory. They may, if they wish, grant them something—the "gifts" Taney refers to—but the theory itself guarantees that the most those outside the system can obtain are gifts. They are not entitled to claim anything because they have no standing within the theory.

So, if the frozen embryos have moral standing, that is no small matter: They, or others for them, can make claims to which those empowered morally must, morally, pay attention. But the determination of what entities have standing cannot be determined independently of commitment to a moral theory. Determining which entities have standing is a subset of the problem of determining the moral scope of a theory—the entities its principles range over and give moral value to, and the entities that are the foci of moral relations according to the theory—what exhaust the moral universe articulated by the theory. This problem of determining a theory's moral scope and, more specifically, the moral entities within a theory entitled to standing, though, is a quite general problem that cannot be resolved, I shall argue, independently of commitment to a theory: A theory and its scope are entwined, a commitment to one determining the other, and vice versa.

Moral Theories and Standing

Consider Rawls' theory of justice. It is justified by determining what those in an original position of equality would choose under such relevant constraints as moderate scarcity. It is a feature of the structure of that justificatory scheme that those who have standing to claim justice under his theory of justice are those in the original position: Those beings who have made it into the original position choose a theory of justice *for themselves.* So, what I call the *entry conditions* into the original position determine which entities have a claim to justice and which do not. We can get a sense of how these entry conditions concern us by asking, "What would original contractors say about such entities as frozen embryos?"

The answer will depend on whether or not those in the original position could end up as frozen embryos when they proceed from the original position to a social position. If they could, one concern would be that there be no more discrimination regarding frozen embryos than color or sex. Being a frozen embryo would be a morally arbitrary contingency and ought to make no more a difference to how one is treated in a just society than whether one is black or white, male or female.

But, if the original contractors could not end up being frozen embryos, any moral concern they might have about such entities would come from a consideration not of justice, but of what would be morally preferable *for them,* that is, for those within the contractual position. Any moral status the frozen embryos would have would be a gift from the contractors whose relations with one another are regulated by a theory of justice—a gift to be given out of the self-interest, even if enlightened, of those contractors and a gift that could be taken away.

The original position is a bargaining position, with self-interested contractors, disinterested in the interests of others, striving to obtain as much as they can for themselves in the way of basic goods essential to achieve their own conception of the good life. They are thought to be behind a veil of ignorance that precludes them from knowing either their social circumstances or natural features, and they are concerned that a bad choice may condemn them to a life of misery should they end up with the wrong contingent features in a society when they have chosen a biased theory of justice. These various features of the original position compel these contractors to choose a theory that guarantees equality to them, unless inequality makes all better off. They bargain to get the best deal for themselves, and the conditions of the original position force them to choose Rawls' theory of justice or, under special conditions, his two principles of justice.[37] Beings who are not originally within the original position do not take part in the bargaining. They are not, therefore, entitled to justice: They are not contractors and can make no just claim on any contractor. Their moral status will be fixed partly by their not being in the original position and partly by what is in the interests of those who are in the original position.

Clearly, it matters morally who gets into the original position, who counts as a contractor to begin with. The question of who contracts is the question of what entities are entitled to justice. So, the entry conditions into the original position have immense moral implications and are, to that extent, not morally neutral: It matters morally what features are picked out as relevant to allow entry.

If an entity is not an original contractor, then that entity is not only not entitled to justice, but, because of the constraints imposed on the original condition, cannot obtain justice from the original contractors. To obtain justice would, in contract terms, amount to being made an original contractor: One becomes a peer with those who choose a theory of justice if one is entitled to what one obtains and it is not just a gift. Those who are contracting, though, have no reason whatsoever to let anyone else contract and good reasons not to let any other entities contract. No contractor acting out of self-interest and disinterest in the interests of others, faced with a moderate scarcity of basic goods, would rationally choose to expand the number of beings who are to obtain a share of such goods. The very way in which Rawls has set up the decision procedure and, in particular, the constraints regarding the self-interested motivations he has imposed on those making decisions, makes it unreasonable for contractors to make any who are not original contractors, members of the original position, covered by the theory of justice.

So, the question of whether an entity is entitled to justice is the question of whether an entity is entitled to enter into the original position. Yet the determination of who gets in cannot be settled by appealing to the original position. If such entities are within the original position and faced with the question whether they should be, they will answer in the affirmative—out of self-interest and a disinterest in the interests of others. And if they are not within the original position, then those within will not let them in. More generally, one cannot appeal to contract theory to settle whether such entities ought to be entitled to justice, because it is their standing to be entitled that is at issue, and the structure of contract theory requires a prior determination of who has standing so as to populate the original position to choose a theory of justice. One cannot use the original position to determine who should populate it without either presupposing that the entities in question have standing, and so can enter, or presupposing that they are not in the original position originally and so guaranteeing that they will not get in.

Rawls limits those who can enter to moral persons, and the justification is a contractual condition, namely, the capacity to keep one's contracts.[38] This contractual condition is not determined by appeal to the original position, but by an appeal to what makes such a contractual position possible: Only if those who contract are capable of keeping their contracts, no matter how they may fare in the final distribution of social circumstances and natural features, will the choice of a theory of justice settle matters and, thus, provide a firm basis for a just society.

This is not to say that Rawls' entry condition is itself morally acceptable, and it is not to say that it does not beg any questions. After all, the structure of Rawls' justification determines that the most relevant feature of those entitled to justice is that they be capable of keeping their contracts, that it is beings capable of such a sophisticated response to the lottery of life who are entitled to a fair share. So, his entry condition is not without its problems, but its problems are not of concern here. What is of concern is that the entry conditions determine who is entitled to justice and who is not and that the justificatory structure of Rawls' theory makes the determination of standing to demand justice a function of the theory itself: Those who have standing are peers entitled to choose a theory of justice that gives them equality unless inequality is to the benefit of all of them.[39]

One can run through the same scenario regarding utilitarianism. Mill claims that the utility principle applies to "the whole sentient creation." Any being capable of feeling pleasure and pain must be considered in any calculation of utility regarding any issue that may affect that being. Even chipmunks have standing in such a utilitarian world. But Mill often treats moral questions as though they concerned only human beings. Yet one will get different moral answers, and have different moral concerns, depending on what beings have standing in a utilitarian world. If animals are left out, the question of vegetarianism will be settled by appealing to the pleasure and pain of humans alone; if animals have standing, their pleasures and pains must also be considered. It is, to put it mildly, not obvious that one will get the same answer on both calculations. Who has standing thus matters morally, and it is arguable that by Mill's theory we ought to consider all beings capable of pleasure and pain in making moral judgments.

Standing under Mill's utility principle is determined by the commitment to take as morally relevant only pleasure and pain: Any being capable of pleasure and pain has a moral stake in a moral universe that seeks to maximize pleasure and minimize pain.[40] One cannot commit oneself to those beings having standing without treating as morally relevant what those beings have in common that makes them subject to the theory and is made morally relevant by the theory. The reason the capacity to keep one's contracts is an entry condition into Rawlsian contract theory and is, thus, the morally relevant feature of all those beings that could be considered as moral beings for Rawls is that Rawls' theory is a contract theory: The structure of justification for that theory guarantees what beings have standing within it.[41] Just so, the determination that pleasure and pain are the only characteristics that are morally relevant determines standing within Mill's theory.

But that determination is not made by using the utilitarian principle. One cannot use the utilitarian principle to determine, for instance, whether one ought to consider the pain and pleasure of animals in deciding to be a vegetarian or only the pleasure and pain of human beings. One will get different results depending on who is supposed to have standing within the principle, and one cannot use the principle itself to determine who has standing. One needs some independent criterion.

That criterion is provided by the theory itself. Just as Rawls' theory takes as morally relevant the capacity to keep one's commitments, so Mill's theory takes as morally relevant the capacity to feel pleasure and pain—being sentient. One cannot determine standing within a theory by the theory's leading principle because that begs the question. Yet an appeal to a theory's leading principle to settle the question of standing begs the question only because that principle articulates the theory's prior commitment to a particular set of entities having standing.

Mill's theory and Rawls' thus categorize the world in different ways: What counts as a moral entity in one's world may not count, and certainly will not count in the same way, in the other's. The theories pick out different features as having moral relevance.

So, one implication is that each theory will beg the question against the other. For instance, Rawls' theory begs the question against utilitarian theories of justice. After all, if all of sentient creation were in the original position, it is not obvious that any being there would choose Rawls' theory of justice. One would not agree that liberty should extend only to persons if one might end up a chipmunk. The contractual condition for entry Rawls insists on skews the decision against a utilitarian concern with all of sentient creation. The dispute between Rawls' theory of justice and utilitarianism cannot be fairly settled from within the original position. The real disagreement, put in Rawlsian contractarian terms, concerns the conditions for entry into the original position.

But what concerns us is that standing within a theory is a function of the theory and not independently determined. To determine that a being has moral standing is to determine to that extent a moral theory: What lets that being in has moral relevance and is given moral relevance by the theory.

Standing in *Davis* v. *Davis*

So, the court's calling the frozen embryos "children, in vitro" is not morally neutral. If that claim means that the embryos have moral standing of any sort, accepting that categorization means accepting a theory in which such entities are entitled to moral consideration. Such a categorization is, thus, freighted with theoretical commitment. For if they have standing in a moral theory, that moral theory must make morally relevant whatever features of the frozen embryos carry them into the moral universe picked out by the theory. So, one cannot make such a decision lightly. It presupposes a moral theory.

If one looks at the frozen embryos and asks of them, "What is their moral status?" it is difficult to know how to respond. It is unclear enough what they are, let alone how one is to respond to them. But if we look at what I call entry conditions into moral theories and consider what moral theories must be like given different entry conditions, we can get a better fix on how we ought to respond, for we get a better understanding of what must be their moral relations and how they are to fare in moral competition, when interests are in conflict. One may view this in the way in which Rawls views the contractual position as an analytical device: Change the conditions of the initial situation, and one gets different choices. So, one way to come to understand different theories of justice is to see the differences in the contractual positions that issue in the different theories.[42] Just so, we can understand more clearly what calling the frozen embryos "children, [even if] in vitro" means, morally, by taking as a sufficient condition for standing in a moral theory whatever natural features of the frozen embryos are supposed to give them standing.

Whatever those features are, the theory that results will be neither Kantian nor utilitarian. Compare standing in Mill's version of utilitarianism with standing in Kant's theory. Kant's theory gives standing to rational beings—God, man, angels. It gives standing, that is, to all creatures capable of using the categorical imperative. Mill's theory gives standing to all of sentient creation—any creatures capable of pleasure and pain and, thus, whose disposition may add to or subtract from the total amount of happiness in the world. Or look at Rawls' theory. The entry condition into Rawlsian justice is a contractual condition, the capacity for a sense of justice. In each of these cases, the natural features picked out as morally relevant are determined by the theoretical commitments. Kant criticizes utilitarians, for instance, for engaging in a sort of high-level anthropology. They investigate what beings of a certain sort, namely, those concerned to further their own pleasure and minimize their pain, ought to do in order to achieve those ends. He rejects the relevance of that natural feature of human beings not by denying that it is a feature, but by pointing out that a utilitarian theory based on it amounts to a set of prudential recommendations and so is not a moral theory at all.[43] He denies, that is, the moral relevance of the only feature of human beings a utilitarian like Mill takes to be morally relevant.

So, what are we to conclude from all this?

1. Moral theories can give us no guidance here, for we cannot appeal to a moral theory to categorize frozen embryos without begging significant questions: Entities have no moral standing independently of a moral theory.

    We might put this point another way. By the nature of the case, a moral theory makes the moral status of some entities problematic. In Mill's theory, it is unclear how we are to count "the whole sentient creation" on a par with human beings. Mill covers the difficulty with the phrase "so far as the nature of things admits,"[44] but the difficulty remains. Similarly, in Kant's theory, it is problematic what we are to do with those entities who are only potentially capable of using the categorical imperative (e.g., fetuses) or those without the potential whom we would regret not having some standing in our moral world and whom we respond to morally (a person claimed to have an IQ of 12). We cannot use the moral theory that makes the moral status of these entities problematic to obtain clarity about their status. Similarly, we cannot use a moral theory to obtain clarity about the moral status of entities the theory excludes from standing.

2. In addition, we should not trust our initial or even considered judgments here. The challenge such entities makes to us is that they do not fit within our moral world in a readily acceptable way. They make otiose the usual moral theories we use to make our way in the world, and so we should not trust our initial or even tutored responses because those may simply reflect the theories these entities challenge. What is required to place properly such entities as frozen embryos in a moral world is a comprehensive moral conception, independently justified, if possible.

3. The entry conditions for frozen embryos into whatever moral theory is required to give them moral standing is considerably weaker than those required by Kant and even the relatively weak conditions required by Mill. The frozen embryos are neither rational beings nor sentient beings: They are not capable of using the categorical imperative, and they do not feel pleasure and pain. So giving them moral standing means that neither Kant's theory nor Mill's theory can be correct. One needs a theory that somehow takes as morally relevant whatever it is that the judge indicates makes the frozen embryos "children, in vitro," and whatever natural features are taken as morally relevant, they cannot be their capacity to feel pleasure and

pain, for they have none, or their capacity for rationality, since they have none. Calling the frozen embryos children is, thus, not morally innocent since it would give us a moral theory in which many more kinds of entities have standing than even in Mill's theory, for, after all, whatever features those frozen embryos have that give them standing, giving them standing will sweep in all other beings analogously placed and all who fall between them and even the minimal condition required by Mill's theory.

So, one implication of giving such entities standing is that that would sweep in fetuses, for surely if frozen embryos have standing, any fetus would. Put another way, the features of frozen embryos most likely to justify giving them moral standing is that they are DNA-coded human and that they have the potential to become persons. But clearly, what separates the frozen embryos from fetuses is that fetuses have crossed a crucial divide in what may be viewed as a process: They are implanted in a woman's womb. So, if frozen embryos have standing because they have the potential to become persons, fetuses surely are better positioned to have standing, being, as it were, a leg up on frozen embryos.

4. To make any suggestion about the moral status of frozen entities is, by my argument, to commit oneself theoretically, and so one ought to make any such suggestion knowing that it will carry in its train much theoretical baggage. So, I make a suggestion with much hesitation. The court seemed frozen by the concepts of the law—children or property—but it surely takes no great leap of legal imagination to conceive of frozen embryos in a different way. They ought to be treated, I would suggest, the way we ought to treat those in a persistent vegetative state, but with a helpful twist. Those in a persistent vegetative state, I would argue, are not persons. Their cerebral cortices are dead or nonfunctioning, and what makes them beings, and

not brain-dead, is that some portions of their lower brain stem are functioning. They are, thus, alive, but alive as something other than the persons they were. They are even, I would argue, not human beings, but some other kinds of beings. We should accord them respect, but because of their history, and their connections with persons in our moral world, not because they have any standing in our moral world. They can neither reason nor feel pleasure or pain or bear any of the moral predicates any entity who is the focus of moral relations would bear: They cannot be responsible, cannot feel guilt or shame, cannot be affected by any decision to minimize pain or maximize pleasure. They have moral status—can be objects of moral relations—but no moral standing.

Frozen embryos are in a similar situation, I suggest. We should accord them respect in part because of their history: They are the result of a complicated and painful process involving two persons who presumably put a great deal of hope and concern into their existence. And so, we accord them respect in part because of their connections to persons in our moral world. But they are not persons: They bear none of the characteristics anyone has ever suggested persons must bear. And if they are alive they are alive as something other than the persons they might become if they were implanted, took, and were carried to term and alive as something other than the animals that Mill would let into his moral world.

But frozen embryos differ from those in a persistent vegetative state in at least one helpful way. We do not yet have, and may not obtain, a test for determining whether the complete shutdown of the cerebral cortex is permanent. We can test to determine if the brain is dead, for we can test to determine if the cerebral cortex and the lower brain stem are functioning, and that test is 100% accurate. But though we seemingly can be as sure as one can reasonably expect in some cases (like that of Karen Anne Quinlan), we cannot be completely sure in all cases. So, though we might commit ourselves to the same sort of general rule about those believed in persistent vegetative states as we have about those who are brain-dead, we cannot apply the rule with the same assurance. But we can be sure that a frozen embryo is a frozen embryo: We need not fret about having criteria that are unclear and so need not worry that our criteria will make it likely that we shall take something other than a frozen embryo and treat it in an inappropriate manner.

I would, thus, suggest that we put frozen embryos into the same sort of category we ought to put those in a persistent vegetative state, with the helpful twist that we can be sure that we have a frozen embryo. The consequence of that certainly is that we need not hesitate to act once a judgment has been made about how such an entity fares when there is a conflict. A frozen embryo would have moral status, but not be the locus of any moral relations—any more than what was a person now in a persistent vegetative state. So, a frozen embryo would have no moral interests and certainly, then, not an interest in being implanted and being born.

Putting frozen embryos into that category does not solve the judge's problem, but does make the conflict clearer. The judge must decide between the interests of Mr. Davis, who presumably does not want children he may have to support financially and may not want children by the woman who was his wife, and the interests of Mrs. Davis, who presumably does want children. It is difficult to know how to respond to such a conflict, but at least it is not distorted by introducing a third focus of interests, those of the frozen embryos. Indeed, if Mrs. Davis should decide not to have the embryos implanted,[45] and if Mr. Davis' interests in not being a father outweigh any other interests of Mrs. Davis, the frozen embryos could be disposed of— inoffensively. As with those in a persistent vegetative stage, the proper test concerns a complex of issues, including the costs of maintaining the entity (the price of a decent shelf-life, as it were), the interests of those persons who are the focus of moral relations involving the entity and the benefits and harms to those interests, the interests of the state in having clear criteria for determining how to respond to such cases so that harm will not occur inappropriately, and so on. Such cases are difficult, but resolvable.

Resolving what theoretical commitments such a view entails, and working out a full moral theory that justifies the sort of standing any theory entails, is also difficult.

Notes and References

[1]References to the case are to *Junior L. Davis v. Mary Sue Davis v. Ray King, M.D., d/b/a. Fertility Center of East Tennessee,* In the Circuit Court for Blount County, Tennessee, at Maryville, Equity Division (Division I), 9-21-89.

[2]*Davis* v. *Davis,* 2.

[3]*Davis* v. *Davis,* 20.

[4]*Davis* v. *Davis,* 17.

[5]*Davis* v. *Davis,* 6.

[6]*Davis* v. *Davis,* 7.

[7]*Davis* v. *Davis,* 15.

[8]*Davis* v. *Davis,* 17.

[9]*Davis* v. *Davis,* A7.

[10]*Davis* v. *Davis,* C1.

[11]*Davis* v. *Davis,* 12.

[12]*Davis* v. *Davis,* 16.

[13]*Davis* v. *Davis,* 15.

[14]*Davis* v. *Davis,* 17.

[15]*Davis* v. *Davis*, 6,20.

[16]*Davis* v. *Davis*, 2.

[17]*Davis* v. *Davis*, 2.

[18]*Davis* v. *Davis*, 20.

[19]*Davis* v. *Davis*, 20.

[20]*Davis* v. *Davis*, A6.

[21]*Davis* v. *Davis*, 12.

[22]*Davis* v. *Davis*, 10.

[23]*Davis* v. *Davis*, 10.

[24]*Davis* v. *Davis*, 10.

[25]*Davis* v. *Davis*, 15.

[26]*Davis* v. *Davis*, 13.

[27]*Davis* v. *Davis*, 17.

[28]*Davis* v. *Davis*, 15–16.

[29]*Davis* v. *Davis*, 15.

[30]*Davis* v. *Davis*, 19,20.

[31]*Davis* v. *Davis*, 20.

[32]*Davis* v. *Davis*, 15.

[33]John Stuart Mill (1957) *Utilitarianism* Bobbs-Merrill, Indianapolis, p. 16.

[34]Ibid.

[35]For the most thorough discussion of this case, *see* Don E. Fehrenbacher (1978) *The Dred Scott Case: Its Significance in American Law and Politics* Oxford University Press, New York.

[36]The Court left open the possibility that Congress might grant citizenship to aliens who were black. Since Congress had not done so, any black living within the United States would not be a citizen.

[37]It is an often ignored feature of Rawls' presentation of his theory of justice that he assumes that the conditions of society in which the contractors will find themselves are improved enough that it is "irrational from the standpoint of the original position to acknowledge a lesser liberty for the sake of greater material gains and amenities of office" (John Rawls (1971) *A Theory of Justice* Harvard University Press, Cambridge, p. 542; *also see* e.g., pp. 151–52). He lifts the veil of ignorance enough, that is, that those contracting know that, however they end up, they will not be willing to trade liberty for any other goods. So, instead of choosing the general theory of justice, in which the goods

are not ranked, the contractors choose the two principles of justice, in which they are.

[38]*See* Rawls, op. cit., especially pp. 505 and 145.

[39]Consider how Rawls must treat the rest of the moral universe. He claims that his contractarian view can be extended to all moral creation, not just to those beings entitled to justice, but those beings who are not entitled to enter into the original position will have their moral status determined by those self-interested beings, disinterested in the interests of others, who do enter the original position. Any other beings have no standing within contract theory unless they are subject to whatever moral principles are chosen from the original position, and, in contract theory, that means they must choose the principles, i.e., must be in the original position. So, to be a moral being entitled to claim justice is to be within the original position. For a more detailed discussion of this issue, *see* Michael S. Pritchard and Wade L. Robison (1981) Justice and the treatment of animals: A critique of Rawls. *Environmental Ethics* **3**, 55–61.

[40]To make my general point, I can ignore such complications of detail as how pleasure is related to happiness and whether Mill is committed to providing the greatest happiness or the greatest pleasure.

[41]Rawls says that moral persons have two characteristics, the capacity for a sense of justice (which implies a capacity to keep one's commitments) and the capacity to have a conception of their good (op. cit., p. 505). I may seem to ignore the latter, but, in fact, I think it is also a contractual condition for Rawls. It is the capacity for a conception of their good that makes moral beings capable of being bargaining agents. So, the capacity for a sense of justice allows moral persons to keep their contracts once they are made, and the capacity for a conception of their good makes moral persons willing to contract to begin with. I have ignored whatever complications this addition would make for the chapter because I do not think it will change the basic claims.

[42]"The procedure of contract theories provides, then, a general analytic method for the comparative study of conceptions of justice. One tries to set out the different conditions embodied in the contractual situation in which their principles would be chosen" (Rawls, op. cit., pp. 121–122).

[43]In the Preface to the *Foundations,* Kant distinguishes between various kinds of knowledge, arguing that though there is what he calls

practical anthropology, the empirical part of ethics, the proper job of philosophy is to concern itself with constructing "a pure moral philosophy which is completely freed from everything which may be only empirical and thus belong to anthropology" ([1985] *Foundations of the Metaphysics of Morals* [Louis White Beck, ed.], Macmillan Publishing Company, New York, p. 5). He goes on to say that "philosophy which mixes pure principles with empirical ones does not deserve the name,...Much less...the name of moral philosophy..." (Ibid., p. 6). I read this as claiming that a utilitarian theory, which bases its principles on a natural feature of human beings, is not really philosophy at all—the philosopher's ultimate rejection of a competing theory.

[44]Op. cit., p. 16.

[45]I heard a news report that she does not want them implanted now, but I have been unable to verify the report.

Creating Children
to Save Siblings' Lives

A Case Study for Kantian Ethics

David W. Drebushenko

Life quickly got more complicated for the Ayala family when, in the Spring of 1990, they drew the attention of the popular and semitechnical press.[1] Only two years before, Abe and Mary Ayalas' 17-year-old daughter, Anissa, was diagnosed with chronic myelogenous leukemia, a disease that rarely afflicts persons under the age of thirty. The only hope for a cure, the Ayalas were told, was to be found in a bone marrow transplant. Anissa's old marrow would first be destroyed by radiation and chemotherapy. New marrow, donated by an individual whose antigens and tissue type were matched to Anissa's, would then be injected through a vein. The search for a donor began. Members of the immediate family were tested and no match was found. More distant relatives of the family were tested, and again, no match was found. Transplant registries operating at the national level were contacted and they failed to identify a suitable donor. It seemed there was no other reasonable course of action left to try.

Mary Ayala had another idea. She would undertake to conceive another child by her husband, Abe, with the explicit inten-

From: *Biomedical Ethics Reviews • 1991*
Eds.: J. Humber & R. Almeder ©1991 The Humana Press Inc., Totowa, NJ

tion of using bone marrow cells donated by her infant to save Anissa's life.[2] At 43, she had a 73% chance of conceiving. The fact that Abe Ayala would have to undergo a procedure to reverse a vasectomy performed 16 years earlier produced another complication. The chance that the reversed vasectomy would lead to a pregnancy is estimated to be 50%. The chance that siblings of the same parents would have matching bone marrow is 25%. And, although the chance that a bone marrow transplant would cure Anissa's leukemia is reasonably high at 70%, a simple theorem of the probability calculus reveals that the overall chances of success are a mere 6%. Given these meager odds, one would not be all that wide of the mark to regard Mary Ayala's plan as desperate.

It would seem the Ayalas cannot be faulted morally.[3] Their effort to save their daughter's life, a morally laudable purpose considered in and of itself, can be regarded as heroic, even if others are inclined to view it as utterly desperate. Pregnancies undertaken at Ms. Ayala's age are standardly treated as high-risk. There is the prospect of sustained discomfort during pregnancy and severe physical pain during delivery. The family claims to want the child for its own sake and not solely for what it can provide to Anissa. Clearly, their actions involve deep emotional attachments and continuing responsibilities. The total burden, measured in physical, emotional, and economic terms, that the Ayala family is willing to assume is inestimable. That they would even seriously consider such a sacrifice is a testament to their devotion to Anissa. That they would willingly endure it testifies to their strength and character.

There are, however, some persons who think the Ayalas are morally wrong. In this case, as in many others like it, one expects opinions to diverge. The divergence, one would have thought, wouldn't be simple; one would not expect there to be just two sides, but many. There is, though, a clear consensus of opinion (some of it surprisingly untentative, in my view) that the action taken by the Ayalas is morally wrong. I disagree. This chapter will attempt to explain why.

Treating Others as Means

One might think that the moral philosophy of Immanuel Kant could be used to underwrite an argument against the Ayalas.[4] Some medical ethicists who have spoken against the Ayalas have alluded to Kant's view that we should never treat others as a means to an end.[5] This principle, or something akin to it, can be found in Kant's "Formula of the End in Itself." On Kant's view, persons, or more generally, rational agents, exist as "ends in themselves." Whether something exists as an end in itself is deeply tied to its rational nature, if it has one, in ways that Kant isn't always clear about. Kant is trying to develop universal, practical principles that set limits on the will, and hence, on human action. The universality, as well as the objectivity of such principles, depends on a notion of absolute value. One of Kant's key ideas is that, "...Rational nature exists as an end in itself." What exists as an end in itself has absolute value. Because of their rational natures, persons, it can be said, exist as ends in themselves. Consequently, persons have absolute value.

Kant's injunction to always avoid treating others as a means to an end is developed from his idea that persons have absolute value. Kant puts some of these ideas in the following way:

> Persons, therefore, are not merely subjective ends whose existence as an object of our actions has a value for us: they are *objective ends*—that is, things whose existence is in itself an end, and indeed an end such that in its place we can put no other end to which they should serve *simply* as means; for unless this is so, nothing at all of *absolute* value would be found anywhere.[6]

In this passage, Kant makes two claims: (1) persons are beings whose "existence is in itself an end," and (2) there is "no other end to which they should serve *simply* as means," which can be substituted for a person's "objective end." However, it is not clear what the premises of Kant's argument are supposed to be.

He certainly thinks there is absolute value or, at the very least, that absolute value is a serviceable notion. What is unclear is whether he means that: (a) there is no absolute value unless persons have an objective end, or (b) there is no absolute value unless there is no end for which persons should serve simply as means that can be substituted for their objective end. It is possible to resolve this difficulty by taking Kant's second claim as an elaboration on, and further development of, his notion of an objective end. An objective end is, at least in part, an end that cannot be replaced by an end for which a person should serve simply as means. On this reading, the premise relevant to Kant's argument is the first. So, since there is absolute value, it follows for Kant that persons have an objective end. And since persons have an objective end, it follows for Kant that no other end for which they serve simply as means can replace their objective end.

How then does Kant "derive" the view that we should strive to avoid treating others simply as means?[7] This is a tricky and complex question. Virtually any answer to it is going to be problematic. My version of the question contains the locution, "(How) does Kant 'derive' the view..." An answer to this question will quickly encounter problems connected with Kant exposition and scholarship. There is, however, an interesting variant on the question by which, arguably, a correct answer will be difficult to come by. It is this: How can the view that we should not treat others simply as means be derived from Kant's metaphysics of morals? Suppose Kant is largely correct in his moral metaphysic; that is, suppose he is right that persons have objective ends, there is absolute value, and so on. His principle that we ought to avoid treating others simply as means has prescriptive content; it is a principle that is intended to guide the actions of a rational agent. How, to put the question briefly, can a prescriptive claim be derived from a metaphysical claim or claims?

Fortunately, this problem does not have to be resolved here. Enough of Kant's moral philosophy has been produced here to make it possible to sketch out a Kantian argument against the

Ayalas. Having done so, I want now to provide that argument. I cannot say whether any commentator on the Ayala case would endorse the details of the argument I'm going to sketch. However, I am reasonably confident that those who would argue against the Ayalas along Kantian lines have in mind an argument whose general features are present in the one I'm going to give.

I take it, then, that a Kantian argument aimed at showing that the actions undertaken by the Ayalas are impermissible might go as follows: A newborn or an infant old enough to donate living tissue is a being whose existence is an end in itself; such a being has an objective end. It is wrong to treat such beings merely as a means to an end. Conceiving a child with a view toward harvesting its bone marrow cells is treating it as a means to an end. Using the child's bone marrow cells to treat a sibling's leukemia is a means of accomplishing the end of saving the afflicted sibling's life. Trying to save the life of a young woman is certainly a morally praiseworthy end or objective. However, on Kant's view, it is of little consequence that the end is morally laudable. What is morally relevant is whether the child is being treated merely as a means. That the child is being treated (or regarded) as a means is established by the fact that its parents intend to use its bone marrow in the pursuit of an end or goal. That the child is being treated merely as a means is established by the fact that its value is derived solely from its usefulness in saving its sister's life. So, it is morally wrong to conceive a child with a view toward using its bone marrow to save its sister's life, since doing so would be to treat the child merely as a means to an end.

One might be inclined to question whether Kant's "supreme principle" is applicable to a newborn or even an infant of 18 months. I began the argument with the claim that a newborn or young infant is a being whose existence is an end in itself. Kant's principle sets limits on how we can act toward beings whose existence is an end in itself. However, as I explained, whether a being has an objective end depends on whether it has absolute value. Whether it has absolute value, in turn, depends on whether it has a "rational nature." At best, newborns and infants of 18 months are "dispositionally rational." Under normal circumstances, such beings will exhibit rationality. Now, there may be a problem here. If having a rational nature requires that the subject be "occurrently rational," then newborns and young infants will not fall under the scope of Kant's principle, since newborns and young infants are not occurrently rational. On the other hand, if by having a rational nature Kant means dispositionally rational, then newborns and young infants, it would seem, fall within the scope of Kant's imperative. Such beings, because it is of their nature, have the disposition to be rational and will, at some point, be occurrently rational.

Even if there is a problem of the type I have been discussing, I would be inclined to overlook it. For the argument can be faulted for slightly subtler reasons. The Kantian argument against the Ayala's that I gave goes wrong in the contention that the only purpose in conceiving the child is to use its cells in an effort to save their daughter's life. The public record makes it plain that the family intends to love the child "...for who she is..."[8] Certainly, there are mixed motives in the case. On the one hand, there is a small chance that the strategy will work; there's a chance that Anissa's life can be saved by the gambit. This fact furnishes a reason for wanting the child and even some basis for using it. However, since the objective probability of success is so slim, and this fact is known to the Ayalas, it seems unlikely that the major reason or motive for conceiving the child lies in its benefit to Anissa. On the other hand, given, at least, Mary's desire for a third child in any event, and the expressed intention of other family members to love and respect the child for who she is, it seems reasonable to conclude that exploitation is not uppermost in the minds of the Ayalas.

These facets of the case upset the argument based on Kantian principles. In the course of developing his "Formula of the End in Itself," Kant is quite careful in his wording. Before giving full expression to the imperative of the end in itself, he writes:

> Now I say that man, and in general every rational being, *exists* as an end in himself, *not merely as a means* for arbitrary use by this or that will: he must in all his actions, whether they are directed to himself or to other rational beings, always be viewed *at the same time as an end.*[9]

It is particularly interesting to ponder the phrases Kant has chosen for emphasis. The emphasized phrases that caught my eye are, "not merely as a means" and "at the same time as an end." Interestingly, similar phrases appear when Kant states the imperative of the end in itself. He writes:

> The practical imperative will therefore be as follows: *Act in such a way that you always treat humanity, whether in your own person or the person of any other, never simply as a means, but always at the same time as an end.*[10]

These two passages constitute significant textual evidence of how Kant wanted the imperative to be understood. He clearly did not think that the imperative was to be understood simply as an injunction to refrain from treating other rational agents as means. That version of the principle is too coarse. It would, for example, restrain actions aimed at bringing about ends that the person being "used" could have a legitimate interest in achieving. Suppose my sister needs a blood transfusion from my brother. Imagine, if you like, that his blood contains a rare antibody not found in my own blood, or anyone else's, that my sister requires in order to fight off a potentially fatal infection. Imagine further that I and my brother wish our sister to continue to live. Would it be wrong of me to urge or to ask my brother to donate his blood so that our sister might be saved? One would hope not. However, the version of Kant's imperative currently under focus suggests otherwise. In urging that my brother donate his blood, I seek to bring about the end of prolonging my sister's life. It also looks as though I treat my brother as a means to bring about that end. It is, afterall, the antibodies in his blood that will secure the end I (and my brother) seek.

This case would not be a problem for Kant's version of the imperative. It would be morally permissible for me to urge my brother to donate blood even though I do treat him as a means. Kant's principle advises that we should not treat others *simply* as a means, but always *at the same time* as an end.[11] This version of the principle does not rule out treating others as means although it does rule out treating others *simply* as means. If I love and respect my brother for whom he is, or otherwise continue to appreciate or treat him as an end, I can urge him to undergo the transfusion and still be a good Kantian. In urging that he consent to the procedure, perhaps I do regard him, or treat him, as a means to an end, but I do not treat him simply as a means.

By now it is becoming clear how one might argue that the Ayalas do not violate the relevant Kantian principles. The Ayalas have conceived a child in an effort to harvest its bone marrow cells for transplantation to their daughter so that her life might be saved. One can grant that they intend to use the child as a means to an end. However, it does not follow from this fact that they violate the relevant Kantian principles. What has to be shown in order to make a case against the Ayalas on Kantian grounds is that they intend or propose to treat the child simply as a means for accomplishing an end. As we have seen, Kant's doctrine of the "End in Itself" allows for the possibility of treating others as a means. And, as the transfusion example illustrates, there is a good reason to allow for this possibility. Kant's doctrine forbids treating others simply as a means. The Ayalas, though, do not have this in mind. Kant's principle permits them to treat the child as a means provided they also treat it as an end in itself. Now, it isn't all that clear what it means to treat someone as an end, but the fault here lies chiefly with Kant for failing to sharpen the notion.[12] If treating someone as an end means recognizing that he or she has absolute value, then, I submit, the Ayalas have satisfied the proviso.

Treating Others as Ends

I want to conclude by considering whether there is a positive argument on behalf of the Ayalas based on a different version of Kant's principle. I want also to identify two problems from the previous section, and to make a proposal toward solving one of them.

The problems are easy enough to identify: (a) in the context of the Ayala case, it isn't clear that the conceptus or young infant is being treated as a means, or merely as a means just because part of its body is being used to accomplish an end, and (b), in general, it is not clear what it means to treat someone as an end. Consider problem (a). Does it follow that a person is treated as a means if it turns out that he or she has some body part or tissue that is useful in bringing about some end? For example, suppose I am a designer and manufacturer of precision rifle sights, and that I have determined that my wife's hair has exactly the right thickness and tensile strength to make perfect "crosshairs." Each day, without her knowledge and consent, I remove the longer strands of hair from her coat so that more sights can be made. It isn't clear to me that in doing this, I treat her as a means to an end. It is possible to make a distinction between treating a person as a means to an end versus using a person's "property" as a means of accomplishing an end. In the Ayala case, the parents intend to use their infant's bone marrow cells as a means of accomplishing the end of saving their daughter's life. That fact alone, however, does not cinch the claim that they treat the *infant* as a means to an end, for they intend to use the infant's marrow as a means to an end, rather than using the infant as a means to an end.

Even so, I don't think that I am required to solve problem (a), for this problem besets the Kantian argument against the Ayalas that I have criticized, albeit for other reasons. Those who allege that the Ayalas are wrong because they violate the principle of the end in itself must explain why it is that the Ayalas' infant is being treated as a means. Perhaps, the answer is going to be that since the marrow originates with the infant, it too is being treated as a means. The Ayalas are using the infant as a means to the end of obtaining the marrow medically necessary to save their daughter's life. Even if much of this is correct, though, it still isn't clear that a person or "being with a rational nature" is being treated as a means. In truth, the needed bone marrow originates with complex physiological and biochemical processes that occur within the infant's body. Why not say those processes are being exploited as a means rather than the "person" of the infant? I just don't see that exploitation of body parts or bodily processes is *eo ipso* exploitation of a person.

What about problem (b)? Until now, I have been avoiding this difficulty by more or less supposing that one treats another as an end by "recognizing that he or she has absolute value," or, when I discussed the transfusion case, I suggested that treating my brother as an end is merely a matter of showing "love and respect for my brother for whom he is." Nevertheless, it isn't clear to me that doing either of these is a sufficient condition for treating someone as an end. Arguably, Kant's clearest application of the imperative of the end in itself occurs in his discussion of lying or false promises. Some of what he says in this connection can be used to address problem (b). He writes:

> ...the man who has a mind to make a false promise to others will see at once that he is intending to make use of another man *merely as a means* to an end he does not share. For the man whom I seek to use for my own purposes by such a promise cannot possibly agree with my way of behaving to him, and so, cannot himself share the end of the action. ...it is manifest that a violator of the rights of man intends to use the person of others merely as a means without taking into consideration that, as rational beings, they ought always at the same time be rated as ends—that is, only as beings who must themselves be able to share in the end of the very same action.[13]

Kant seems to have felt that lying promises are morally troublesome because in making one, you treat the person of another merely as a means to an end the other person does not and, according to Kant, could not share. This passage is interesting because (among other things) it contains a suggestion for dealing with the problem of interpreting Kant's locution of "treating others as ends." When Kant says that rational beings "ought always at the same time be rated as ends," he goes on to add that such beings "must themselves be able to share in the end of the very same action." The idea is that one will act in such a way as to treat others as ends provided others "share" in the end brought about by the action under consideration. The transfusion example works as a nice illustration of the point. Recall that in putting forward the example, I had assumed that my brother and I wanted our sister to continue to live. It is true that the antibodies in my brother's blood are being exploited in an effort to keep our sister alive. And, *perhaps* because of this, it may be said that I treat my brother as a means to an end in urging him to have the transfusion. Even so, I do not treat him merely as a means to an end in asking him to undergo a procedure that will save our sister's life, since he has an interest in seeing that she lives.

The suggestion, then, is that Kant's talk of treating others as ends can be explicated in terms of their having an interest in accomplishing the relevant end. However, this way of putting the suggestion does not quite square with Kant's word. For, on this formulation of the suggestion, one will treat others as ends provided they actually have an interest in accomplishing the relevant end. What Kant says inclines one to think that others are treated as ends if "they are able to share in the end," so that what is crucially relevant to the treatment of others as an end is that it be possible for them to share in the end. Notice also that Kant uses the word "share." Thus, I don't think it is altogether accurate to say that others are treated as ends by virtue of having a mere interest in bringing about a given end. Kant's term is ambiguous. It can mean (1) others may be said to share in an end if it is a common goal, or (2) others can be said to share an end if they profit from its production. Although I cannot say which meaning Kant intends, I don't think this presents much of a problem, for on many occasions, both senses of "share" will surface. In the typical case, one expects that persons have common goals because they will benefit from reaching those goals. And, typically, if all the parties to an action stand to benefit from its end, then they will have a common goal.

It is now possible to sketch out an argument that seeks to exonerate the Ayalas on (modified) Kantian grounds. Let us suppose that Kant's principle is to be understood as follows: Never treat a being with a rational nature merely as means to an end, but always at the same time as a being who can share in the relevant end. The Ayalas intend to conceive a child so that its bone marrow can be used to save its sister's life. Clearly, it is possible for the child to share in the end of saving its sister's life. There is a good bet that at some future time the child will benefit by having its sister's life preserved. So, even if the Ayalas intend to treat the infant as a means, they do not treat it merely as a means. Consequently, the Ayalas act in accord with Kant's (modified) principle.

Notes and References

[1] *See Time,* March, 5, 1990, vol. 135; *People Weekly,* March 5, 1990, vol. 33; and *American Medical News,* March 2, 1990, vol. 33.

[2] Discussions of similar kinds of cases have already appeared in the medical ethics literature. *See,* for example, Alan Fine (1988) The ethics of fetal tissue transplants, *Hastings Center Report* (June/July), **18,** 5–8. *See also* John Robertson (1988) Rights, symbolism and public policy in fetal tissue transplants, *Hastings Center Report* (December), **18,** 5–12.

[3] There is at least one writer I know of who would probably agree with this assessment, although for wildly different reasons. *See* Mary Ann Warren's piece in Mary Ann Warren, Daniel Maguire, and Carol Levine (1978) , Can the fetus be an organ farm? *Hastings Center Report,* (October), **8,** 23–25.

[4]There are other discussions of the application of Kantian principles to problems in medical ethics. *See,* for just one example, J. W. Walters and Stephen Ashwal (1988) Organ prolongation in anencephalic infants: Ethical and medical concerns, *Hastings Center Report* (October/ November), **18,** 19–27.

[5]*See,* for example, the comments by George Annas and Richard McCormick (1990) in *American Medical News,* (March 2), **33,** 3,9.

[6]This passage is taken from Kant's (1964) *Groundwork of the Metaphysics of Morals* (H. J. Paton, eds.), Harper and Row, NY, p. 96. The itallics are Kant's. Hereafter, I shall refer to this piece as *The Groundwork.*

[7]For more on the complicated issue of "derivability," the interested reader might wish to consult Bruce Aune (1979) *Kant's Theory of Morals,* Princeton University Press, Princeton, pp. 70–77.

[8]*People Weekly,* March 5, 1990, vol. 33, vol. 135. *See also Time,* March 5, 1990.

[9]*The Groundwork,* p. 95. All emphases are Kant's.

[10]The Groundwork, p. 96. The emphasis is Kant's.

[11]Incidentally, H. J. Paton, in his translation of *The Groundwork,* observes in a footnote that "... 'simply' is essential to Kant's meaning since we all have to use other men as means." *The Groundwork,* p. 139. *See also* Paton (1958) *The Categorical Imperative: A Study in Kant's Moral Philosophy,* Hutchinson, London, p. 165. For discussion of Paton's view of the imperative, *see The Concept of the Categorical Imperative* (1968), T. C. Williams, Oxford University Press, London, pp. 67-79.

[12]I don't think it is altogether fair to blame Kant for the unclarity. As I explain in the next section of the chapter, there are some things Kant does say that can be used to solve this problem.

[13]*The Groundwork,* p. 97. The emphasis is Kant's.

Fetal Tissue
Transplantation

An Update

Mary B. Mahowald

Politically, the issue is moribund, and is likely to remain so unless and until the political climate changes.[1] During the past five years, advances in techniques for fetal tissue transplantation have raised many people's hopes of finding cures for previously incurable and devastating diseases. Simultaneously, these advances have provided a new target for opponents of legalized abortion.

Background

In 1986, the author of this article responded to the concerns of a colleague, Dr. Jerry Silver, who was working on the problem of spinal cord regeneration in mice, by convening a national meeting of neuroscientists, ethicists, and legal scholars to address the issue of fetal tissue transplantation. Dr. Silver was concerned about whether successful efforts by fetal "transplanters" working with primates might be permitted to proceed towards

From: *Biomedical Ethics Reviews • 1991*
Eds.: J. Humber & R. Almeder ©1991 The Humana Press Inc., Totowa, NJ

relevant human applications, or whether the association between abortion and abortion as the means through which such human tissue becomes available would preclude that progress.[2]

The aim of the forum held in Cleveland, Ohio, on December 4 and 5, 1986, was to gather the input of those most knowledgable about the issue, and to share this with the public in order to precipitate an informed debate on a prospective rather than retroactive basis. To this end, in addition to participation by appropriate experts, representatives from the National Institutes of Health, the lay public, and local practitioners with pertinent expertise were also invited. Following broad discussion with the participants, the presenters formulated a consensus statement that was published in *Science* on March 13, 1987.[3] Their revised presentations were published in *Clinical Research* in April 1988.[4] Despite this early impetus to address ethical parameters of the issue proactively, in March 1988 the federal government imposed a moratorium on use of human fetal tissue from induced abortions for transplantation.[5] During the fall of 1988, a panel of experts convened by the National Institutes of Health met to examine the issue; their report was published in December, 1988.[6] Three of the 21 members of the panel dissented from its concluding recommendations to the Assistant Secretary of Health. The major recommendations, consistent with those already published by the Cleveland forum, were the following:

1. The timing of abortions must not be linked with the research use of fetal tissue.
2. A woman's decision to donate fetal tissue remains is a "sufficient condition for their use" (that is, under most circumstances, no other consent is necessary).
3. "Profiteering" in, or "valuable consideration" for donations of fetal tissue, should be strictly prohibited.
4. General knowledge of potential biomedical research uses for fetal tissue should not be considered a "prohibited inducement" for a woman's decision to have an abortion.

5. Because current research leading to fetal tissue transplant experiments in humans is showing promise, the ethical questions it poses should be addressed now.[7]

A substantial majority of the NIH panel thus supported use of fetal tissue for transplantation, under conditions intended to prevent ethical abuses. Nonetheless, in the fall of 1990, the government moratorium was extended indefinitely.[8] At this writing, therefore, although fetal tissue transplants may legally be performed in most states, the federal funding on which most scientists rely for their support is denied to those whose research involves fetal tissue obtained from legal abortions. The work of American transplant researchers is thereby impeded, while their counterparts in other countries continue to make progress. Political pressures not to support the conclusions of its own panel have won the day in the United States—at least for the present. The impasse between the government's decision and the recommendations of its advisers represents an ethical dilemma in its own right.[9]

Despite the relative novelty of the issue in people's consciousness, it is not, in fact, a new one. Use of fetal tissue for transplantation has been performed in animals at least since the turn of the century.[10] In humans, transplantation of fetal tissue from human as well as nonhuman animals has occurred with little controversy for many years.[11] Apparently, it was the possibility of using *neural* fetal tissue that first provoked public debate about the technique, and this led to concerns about the means through which the tissue would be obtained, namely, abortion. Because abortion is also the means through which nonneural fetal tissue is obtained, those who judge abortion as immoral and use of tissue obtained therefrom as complicity in its immorality might have challenged the use of nonneural fetal tissue as early as 1968. In that year, human fetal thymus became the therapy of choice for treatment of DiGeorge's syndrome.[12] In what follows, therefore, I will consider two aspects of fetal tissue transplantation that have emerged as central in the debate: the use of *neural* tissue, and the association between abortion and fetal tissue transplantation. Different paradigms or frameworks for determining the ethics of fetal tissue transplantation will also be discussed.

Transplantation Involving Neural Tissue

Neural grafting is a surgical procedure for transplanting tissue from various sources into specific areas of the nervous system that have been affected by a neurologic disorder, disease, or injury.[13] Sites from which fetal tissue has been retrieved for neural grafting include the brain and the adrenal gland of the fetus. Although most experimental work in neural grafting has involved transplantation into the recipient's brain, additional sites include the spinal cord and the peripheral nervous system.[14] Neurological disorders that are potentially curable or remediable through fetal grafts include Parkinson's disease, Huntington's disease, Alzheimer's disease, amyotrophic lateral sclerosis (Lou Gehrig's disease), multiple sclerosis, epilepsy, brain or spinal cord injury, and stroke.[15]

Adrenal fetal transplants into the recipient's brain, and transplants from or into parts of the body other than the brain (e.g., fetal thymus, liver, pancreas) are ethically less controversial than transplants of fetal brain tissue into the recipient's brain. Unlike other organs, the brain is prevalently identified with an individual's distinct personality. To the extent that tissue removed from the fetal brain represents a distinctly different personality than that of the recipient, the problem of identity arises. Would the recipient thereby assume a different (or perhaps an additional) personality? Presumably, the import of the identity problem depends on the proportionate amount of tissue transferred and the degree of development achieved by the fetal brain. The size of a fetal brain is of course much smaller than the brain of an adult. Just how much fetal brain tissue should be transplanted for optimal results in the recipient has not yet been determined. Theoretically, it is possible for a minute portion of brain tissue to be removed stereotactically from a living donor and implanted into the brain of a recipient, without seriously jeopardizing the health or identity of the donor.[16] It may thus be possible to use so minute a portion of fetal brain tissue that the question of recipient identity is trivialized.

The problem of identity is also undercut by the fact that the fetal brain is relatively undifferentiated in comparison with the recipient's brain. Obviously, the earlier the gestation, the more undifferentiated the fetal tissue. But neither the proportionate amount nor the degree of development of fetal brain tissue affects the recipient's identity unless the brain truly is the source of that identity. Because scientific evidence is apparently unable to offer an empirical explanation of the relationship between personal identity and the brain, it remains a matter of philosophic debate. Different concepts of a nonphysical "soul," "mind," "person," "individual self," and its relation to the physical brain give rise to various theories, all of which are subject to critique and controversy.[17]

Even if it were established that the brain truly is the source of personal identity, the relevance of this view to fetal tissue transplantation must be weighed against the probability that the recipient's life would otherwise be cut short or lost. Thus, the argument: better to live even with another's (partial) identity than not to live at all. Life is the more fundamental value, on which identity itself depends.

In contrast with brain grafts, neural grafts from or into the spinal cord or peripheral nervous system do not evoke concerns about personal identity. At present, fetal tissue transplantation for treatment of severe neurologic disorders in humans is experimental because its efficacy is not confirmed, even with regard to the disease for which research has been most promising, namely, Parkinson's disease.[18] Treatment of neurological disease through transplantation of fetal adrenal tissue is largely unsuccessful.[19] Treatment with fetal neural cells has been preliminarily successful in only a few centers.[20] Nonetheless, hope runs high that further animal studies as well as work with humans will lead eventually to standardization of this mode of treatment.[21]

Abortion and Fetal Tissue Transplants

Although brain cell transplants may have sparked current interest in the issue, the relation between fetal tissue transplantation and abortion has sustained the interest, evoking some of the same passion and sloppy rhetoric as has surrounded the abortion issue all along. Just as abortion has been compared with a silent holocaust, so those supporting use of electively aborted fetuses for transplantation have been accused of fomenting another holocaust.[22] Proponents of a legal right to abortion see little problem in using fetal tissue for research or therapeutic purposes. Those who argue that the fetus is a woman's property are (or logically ought to be) open to commercialization of fetal tissue, so long as women receive their just profit from such sales.[23] It may even be argued, following Ruth Macklin's suggestion about research, that not using aborted fetal tissue for transplantation is morally irresponsible.[24]

In general, not only "right to life" activists but also those who support women's right to abortion view the choice as a tragic option. It interrupts a natural human process, with concomitant physical discomfort, and loss of a potential human being. These negative aspects may be outweighed by expected benefits but they are negative nonetheless. Hence, most people would not like to see the practice of abortion encouraged by the kind of inducement that the prospect of fetal tissue transplantation represents.[25] It may be highly unlikely, as John Robertson has argued, that the possibility of donating fetal tissue "will contribute significantly to the rate of abortion,"[26] but even minimal increase is a factor to be weighed in formulating public policy. Accordingly, the possibility of obtaining the desired tissue through other means is worth pursuing. Spontaneous abortions, ectopic pregnancies, and fetal tissue culture have all been recommended as ways through which to bypass the problem of elective abortion.

The first instance of Parkinson's treatment through use of human fetal tissue occurred in Mexico, where abortion law is highly restrictive. Spontaneous abortion was predetermined as the means through which the tissue would be acquired.[27] Although the investigators reported that their technique had been preliminarily successful, serious questions have been raised by others regarding their work.[28] Most view spontaneous abortion as an inappropriate source of tissue because it is more likely to be defective than tissue obtained from elective abortions. An unacceptable risk for the recipient is thus introduced. Moreover, because spontaneous abortions usually occur outside of hospitals, the likelihood of retrieving viable tissue by this means is not very high.[29]

Kathleen Nolan has proposed that ectopic pregnancies be considered as a source of fetal tissue for transplantation.[30] Over 75,000 ectopic pregancies occur each year, and the fetal tissue removed through surgery for this life-threatening condition is unlikely to have substantially increased incidence of chromosomal abnormality. From an ethical point of view, surgical removal of an ectopic pregnancy is comparable to therapeutic abortion for maternal health, with the added caveat that the circumstances are already fatal for the fetus. Nonetheless, some of the same problems arise here as with acquisition of tissue obtained through spontaneous abortions. Because ectopic pregnancies require urgent treatment, it is difficult to predict and plan the retrieval of tissue through this means, and the treatment may itself result in damage to fetal or embryonic tissue.

Fetal tissue culture or continuous cell lines developed from a single fetus is an attractive possibility that researchers are presently pursuing. However, this process does not in fact bypass the ethical problem of elective abortion unless the original tissue is acquired through some other means. If the tissue is obtained through elective abortions, fetal tissue culture simply replicates the problem while appearing to be removed from it. Moreover, by perpetuating the existence of fetal cells obtained from the same woman, this approach exacerbates the question of ownership. The recent case of *Moore* v. *Regents of the University of California* suggests the possibility of women suing for profits obtained through use of cultured fetal cells obtained from their abortions.[31]

Other means of treating diseases that are potentially curable through fetal tissue transplants have had only limited success. Autografts of adrenal tissue, for example, although touted at first, have not fulfilled their promise, and conventional means of treatment, such as el-dopa for Parkinson's disease, gradually become ineffective.[32] Fetal tissue transplants have provided an exciting prospect of curing previously incurable, severely debilitating diseases because of the unique properties of fetal cells: their rapid rate of growth and relative lack of differentiation in comparison with more mature cells. In addition, because of the blood brain barrier, neural fetal cells are less likely to be rejected by the recipient than other donor cells.[33]

On therapeutic grounds alone, a comparison of the potential advantages of using fetal tissue from electively aborted fetuses with the potential and actual disadvantages of treatment through other means provides a strong case for use of fetal tissue from elective abortions. However, therapeutic efficacy alone doesn't constitute moral justification. This returns us, then, to the question of whether elective abortion is morally separable from fetal tissue transplantation. The issue calls for reexamination of the traditional moral dilemma involving the relationship between means and ends. Does the end justify the means in transplantation of fetal tissue for cure of otherwise incurable disorders?

A simplistic version of utilitarianism supports an affirmative answer to the question.[34] In other words, the tremendous good that might be accomplished through the new technique outweighs the harm that might be done through elective abortion. However, if endorsement of the procedure leads to widespread increase in elective abortions, a reduced sense of the value of human life, and to exploitation of women, it *is possible* that such an array of undesirable consequences would outweigh the potential benefit of the technique. So, even if the end justifies the means, it is not clear that it does so in this case. Whether or not the overall consequences of treating debilitating disorders through fetal tissue transplantation will generally constitute a preponderance of harms over benefits is an empirical issue for which more data is needed to support a credible utilitarian position.

From a deontological point of view, the end does not justify the means, but this does not necessarily imply that fetal tissue transplantation is morally unjustified. The individual who knowingly and freely pursues a specific end, also, knowingly and freely chooses the means to its fulfillment. In other words, intention is crucial to the moral relevance of the relationship. If one were to deliberately become pregnant, choose abortion or persuade another to do so solely for the sake of fetal tissue transplantation, one would then be responsible for both means and end because one would be intending both. Along with that intention, the motive of the decision may be altruistic (for example, treatment of a relative or anonymous patient), self-interested (such as treatment of oneself, or profit through sale of the tissue), or mixed (that is, self-interested *and* altruistic). Although worthy motives are morally relevant at least from a subjective point of view, they do not alter the fact that the intention in such cases applies to both ends and means.

In other situations involving fetal tissue transplantation, the individual who intends to use the tissue in no way intends the abortion through which the tissue becomes available. Presumably, he or she does intend the retrieval procedure. However, just as a transplant surgeon may retrieve essential organs from the brain-dead victim of a drunk-driving accident, without any implication that he or she thus endorses the behavior that led to the availability of the organs, so may a neurosurgeon who is totally opposed to abortion transplant neural tissue from a dead fetus electively aborted into a severely impaired patient without thereby compromising his or her moral convictions. In fact, one may argue that a truly prolife position favors the affirmation of life that the transplantation entails, while acknowledging the negation of life that abortion implies. When the abortion decision has already been made by others, a decision not to transplant seems less in keeping with a prolife position than its opposite.

So go the different arguments for separability of the two issues. But what about arguments against their separability? These are mainly based on concepts of complicity and legitimation.[35] James Bopp and James Burtchaell, who disagreed with the NIH panel's recommended guidelines regarding fetal tissue transplantation, argued that

> whatever the researcher's intentions may be, by entering into an institutionalized partnership with the abortion industry as a supplier of preference, he or she becomes complicit, though after the fact, with the abortions that have expropriated the tissue for his or her purposes.[36]

Those who use fetal tissue from elective abortions thus ally themselves with the "evil" that abortion represents.[37]

Legitimation occurs when individuals considering an abortion—whether pregnant women, their partners, or practitioners—construe the possibility of benefiting someone by donating fetal tissue a positive endorsement of the abortion option. Abortion is then seen as a less tragic choice than it would otherwise be, and in some circumstances it might even be seen as virtuous. Legitimation would occur on a social level if the good of successful treatment through fetal tissue transplant became so compelling that the means of achieving the success were never critically assessed. The end would then have justified the means, at least as perceived by those who pursue the end without scrutinizing the end in its own right.

The legitimation argument illustrates more general concerns about slippery slope reasoning. If, for example, we now approve use of fetal tissue for transplants under restrictive conditions, are we not likely in time to relax the conditions if the therapy proves highly successful or if the restrictive conditions limit its usefulness? Most people agree that some restrictions are necessary to avoid abuses occasioned by the development of the technology; they disagree, however, about where to place wedges along the slippery slope.[38]

Some have proposed less restrictive guidelines than those recommended by the NIH panel, particularly with regard to commercialization. Lori Andrews, for example, argues that a woman should be allowed to sell the tissue of a fetus she has agreed to abort.[39] Feminists, she maintains, are inconsistent with their commitment to promote women's right to control their own bodies if they oppose commercial surrogacy.[40] Most feminists, however, oppose both surrogacy and commerce in fetal tissue on the same grounds, viz., the possibility they present for exploitation of poor women. Unlike Andrews, such feminists place greater emphasis on social equality than individual liberty.[41]

Different views regarding abortion also give rise to different views regarding the consent necessary for fetal tissue transplantation. Those who are morally opposed to the legality of abortion generally deny that women who choose abortion have a right to donate fetal tissue.[42] Such women, they allege, have forfeited that right even as parents may forfeit their right to consent for their child if they abuse or abandon the child. On the other side of the issue are those who stress the importance of the pregnant woman's consent to use of fetal tissue because she has the right to abortion and because the tissue belongs to her.[43] Among those who consider abortion a separate issue from fetal tissue transplantation and who also consider abortion immoral, some insist that the pregnant woman's consent is necessary because the timing and procedure for abortion may be altered in order to maximize the chance for successful transplant.[44] In other words, if the pregnant woman may herself be affected, her consent to use fetal tissue is morally indispensable.

Paradigms and Frameworks

Different paradigms and frameworks have been invoked in order to defend or oppose fetal tissue transplantation. The paradigms include transplantation from living donors, as in kidney transplants; transplantation from cadaver donors, as in heart transplants; and surrogate motherhood.[45] The first two are familiar and generally accepted means of obtaining organs or tissue, so long as consent is obtained from the donor or proxy and the retrieval does not constitute a major threat to the donor's health. Although use of tissue from living fetuses has generally been rejected, it is difficult in some cases to determine whether a fetus is dead. Traditional means of assessing brain death are not applicable to early fetuses or abortuses. Surrogate motherhood is obviously a more controversial practice to invoke as a paradigm for fetal tissue transplantation, but it captures, as the other two paradigms do not, the unique possibilities for exploitation of women that fetal tissue transplantation represents. In both situations, whether or not money is exchanged, women's bodies are used for the sake of another. This suggests analogy with an even less enticing paradigm, prostitution. Class differences like those between surrogates or prostitutes and those who hire them might also occur in fetal tissue transplantation.

The Center for Biomedical Ethics at the University of Minnesota has proposed three "competing frameworks" that may be related to the above paradigms (in the order in which I've mentioned them above).[46] The first is based on the premise that the fetus from which tissue may be retrieved should be regarded as a human research subject. On this view, according to the authors, either of two rationales may prevail, depending on whether the aborted fetus is construed as living or dead. If the former, use of fetal tissue "should satisfy the federal regulations for research involving living fetuses, and be reviewed and approved by an institutional review board."[47] If the aborted fetus is regarded as a cadaver, a proxy decision-maker should be required "to base a decision regarding participation either on the basis of what the dead fetus would have wanted or on some view of what is in the dead fetus' best interests."[48] Not surprisingly, neither of these standards is explained, and the authors acknowledge that it is "extremely difficult to see how a proxy decision-maker could base a decision on [them]."[49]

The second ethical framework proposed by the Minnesota group is a view of the dead fetus as a cadaveric organ donor. This generally means following the standards of the Uniform Anatomical Gift Act (UAGA), which is applicable in all 50 states in the United States. The UAGA allows either parent to consent to use of fetal tissue so long as the other parent does not object. Moreover, because wishes of the dead fetus are unknown (and may in fact be absent), parents may base their decision on their own needs, concerns, and interests.[50] This is the framework utilized in the consensus statement of the 1986 forum in Cleveland. According to the authors,

> retrieval of such [neural] tissue from fetal remains is analogous to the transplantation of organs or tissue obtained from adult human cadavers. Similarities include the fact that the donor is dead, and the expectation that there will be significant benefits for the recipient. These similarities suggest the appropriateness of using the same ethical and legal criteria now followed for cadaver transplantation.[51]

The beginning point of the Cleveland group's, as well as the NIH panel's report, is that fetal tissue should only be retrieved from *dead* fetuses. Only then does the analogy with retrieval of tissue from "adult human cadavers" work. Even then, it is recognized that the differences between transplantation from human fetal cadavers rather than mature human cadavers should be addressed through added requirements, such as the exclusion of familial donors and the observance of anonymity between donors and recipients.

The third framework proposed by the Center for Biomedical Ethics at Minnesota is one in which the dead fetus or abortus is equated with discarded tissue. In that context,

> fetal remains, whether the result of elective abortion, ectopic pregnancy, or spontaneous abortion, are treated as any other bodily tissue and fluid removed during a diagnostic or surgical procedure. [52]

Aborted tissue is thus construed as a tissue specimen of the woman from whose body it was removed. Permission from those whose discarded tissue may be examined for educative, research, or future treatment purposes is routinely obtained in the clinical setting. Typically, the consent forms include "boilerplate" language requesting blanket permission for use of any biological "waste materials" or "tissue specimens" removed during surgical procedures.[53] Similar boilerplate language could be incorporated into the consent form for abortion procedures.

Whereas the first two frameworks proposed by the Minnesota Center focus on the fetus as a separate being from the pregnant woman, the third focuses on the fact that fetal tissue is, in fact, the woman's tissue, and ought to be treated as such even when aborted. It is thus appropriate to ask the pregnant woman for consent to use fetal tissue prior to abortion, and her consent alone is morally adequate. Some might argue that consent of the man who impregnated the woman should also be required for use of fetal tissue, but this suggests an unusual concept of "discarded tissue," and a departure from the usual manner of dealing with discarded tissue. Moreover, so long as abortion is a decision legally made by women and not by their male partners, men cannot effectively challenge pregnant women's decisions regarding disposition of their fetuses.

Like the surrogate motherhood model, the discarded tissue framework emphasizes the essential tie between fetus and pregnant woman. The two are related in that one is a means of avoiding the abuses that we have seen associated with the other. Because the discarded tissue model gives priority to the pregnant woman's autonomy, it serves as a check on the possibilities for exploitation of women that transplantation of fetal tissue allows. There are thus both conceptual and moral reasons for preferring this framework to the others: it takes account of the unique relationship between fetus and pregnant woman, and the practice it engenders is consistent with respect for patient autonomy in comparable situations. As proportionately more women than men become involved in fetal tissue transplantation, the pertinence of the surrogate paradigm and discarded tissue framework will become more evident.

Summary

Although there have been political as well as scientific setbacks in the development of fetal tissue transplantation as a means of treating severe and otherwise incurable diseases, its therapeutic possibilities remain promising. Although philosophically intriguing, the question of whether transplantation involving neural tissue raises special problems about personal identity is less urgent than the question of the relationship between fetal tissue transplants and elective abortion. The majority of those studying the latter question propose a separation of the two issues, recommending that research with fetal tissue proceed under ethical guidelines that preclude a position on the morality of abortion. Paradigms and frameworks invoked to defend this proposal include transplantation from living donors, on which account the fetus may be considered a human research subject; and transplantation from dead donors, on which account the fetus may be regarded as a cadaveric organ donor. Another paradigm is that of surrogate motherhood, which suggests the essential tie between pregnant women and fetuses that may be used for transplantation. Viewing aborted fetuses as tissue discarded by pregnant women also suggests a necessary tie between the two. Different ethical frameworks support different ethical guidelines for use of fetal tissue for transplantation.

Notes and References

[1]US Congress, Office of Technology Assessment (1990) Appendix A: DHHS Moratorium on Human Fetal Tissue Transplantation Research, *Neural Grafting: Repairing the Brain and Spinal Cord*, OTA-BA–462, US Government Printing Office, Washington, DC, pp. 171–173.

[2]This concern led him to join me, along with neurosurgeon Robert Ratcheson, in writing an article on the issue (1987) The ethical options in transplanting fetal tissue, *Hastings Center Report* **17,** 9–15.

[3]Mary B. Mahowald et al. (1987) Transplantation of neural tissue from fetuses, *Science* **235,** 1308–1309.

[4]Mary B. Mahowald et al. (1988) Neural fetal tissue transplantation: Scientific, legal and ethical aspects, *Clinical Research* **36,** 3.

[5]US Congress, Office of Technology Assessment, Appendix A.

[6]Consultants to the Advisory Committee to the Director, National Institutes of Health, *Report of the Human Fetal Tissue Transplantation Research Panel*, December, 1988.

[7]*Ibid.,* Vol. II, A2.

[8]US Congress, Office of Technology Assessment, 149,171.

[9]*Ibid.,* 171.

[10]Alan Fine (1986) Transplantation in the central nervous system, *Scientific American* **255,** 52; Edwin Kiester, Jr. (1986) Spare parts for damaged brains, *Science '86* **7,** 34.

[11]Human fetal tissue was used in the development of polio vaccine during the 1950s, but the first attempts to transplant human fetal tissue occurred as early as the 1920s. Cf. Center for Biomedical Ethics, University of Minnesota (1990) *The Use of Human Fetal Tissue: Scientific, Ethical and Policy Concerns,* (January).

[12]*Ibid.,* 21.

[13]Much of this section is drawn from my article (1989) Neural fetal tissue transplantation: Should we do what we can do? *Neurologic Clinics* **7,** 747–748.

[14]Alan Fine, *op. cit.,* 52–58.

[15]US Congress, Office of Technology Assessment, 93–109.

[16]Mahowald et al., *The Hastings Center Report,* 11.

[17]Cf. Michael Green and Daniel Wikler (1980) Brain death and personal identity, *Philosophy and Public Affairs* **9,** 105–133; and Warren Quinn (1984) Abortion, identity and loss, *Philosophy and Public Affairs* **13,** 24–54.

[18]US Congress, Office of Technology Assessment, 61.

[19]*Ibid.,* 68–71; and Center for Biomedical Ethics, University of Minnesota (1990) *The Use of Human Fetal Tissue: Scientific, Ethical and Policy Concerns,* January, 103–108.

[20]US Congress, Office of Technology Assessment, 61–90; and Center for Biomedical Ethics, University of Minnesota, 109–110.

[21]Cf. US Congress, Office of Technology Assessment, 84.

[22]After publishing the consensus recommendations of the Cleveland forum in Science, (March 19, 1987), I was personally accused in an unsigned letter of fomenting another holocaust. Cf. James Bopp, Jr. and James Burtchaell (1988) Human Fetal Tissue Transplantation Research Panel: Statement of Dissent, Consultants to the Advisory Committee to the Director, National Institutes of Health, *Report of the Human Fetal Tissue Transplantation Research Panel* I, 64–69.

[23]e.g., Lori Andrews (1986) My body, my property, *Hastings Center Report* **16,** 28.

[24]Ruth Macklin (1977) On the ethics of not doing scientific research, *Hastings Center Report* **7,** 11–13.

[25]E. J. Dionne, Jr. (1989) Poll on abortion finds the nation is sharply divided, *New York Times,* (April 26), 1.

[26]John A. Robertson, Rights, Symbolism, and Public Policy in Fetal Tissue Transplants, *Hastings Center Report* 18, 6 (Dec. 1988), 7.

[27]Mary B. Mahowald (1988) Introduction to neural fetal tissue transplantation: Scientific, legal and ethical aspects, *Clinical Research* **36,** 187,188.

[28]Ignacio Madrazo et al. (1988) Transplantation of fetal substantia nigra and adrenal medulla to the caudate nucleus in two patients with

Parkinson's disease, *New England Journal of Medicine* **315**, 51; Center for Biomedical Ethics, University of Minnesota, 109; US Congress, Office of Technology Assessment, 67.

[29]Center for Biomedical Ethics, University of Minnesota, 136–138.

[30]Kathleen Nolan (1988) *Genug ist Genug*: A fetus is not a kidney, *Hastings Center Report* **18**, 18,19.

[31]US Congress, Office of Technology Assessment, 158–161.

[32]*Ibid.*, 95,96; 68–71.

[33]Mahowald ,et al., *The Hastings Center Report*, 10.

[34]Cf. my Neural fetal tissue transplantation: Should we do what we can do? *Neurologic Clinics* **7**, 750,751.

[35]Center for Biomedical Ethics, University of Minnesota, 251–267.

[36]Human Fetal Tissue Transplantation Research Panel: Statement of Dissent, Consultants to the Advisory Committee to the Director, National Institutes of Health, *Report of the Human Fetal Tissue Transplantation Research Panel* **I**, 70.

[37]Robertson, 6.

[38]Mary B. Mahowald (1988) Placing wedges along a slippery slope, *Clinical Research* **36**, 220–222.

[39]Cf. Bopp and Burtchaell, 56.

[40]Lori B. Andrews (1988) Feminism revisited: Fallacies and policies in the surrogacy debate, *Logos* **9**, 81–96.

[41]e.g., Barbara Katz Rothman (1990) Surrogacy: A Question of values, in *Beyond Baby M* (Dianne Bartels et al., eds.), Humana Press, Clifton, NJ, pp. 235–241; Hilde Lindemann Nelson and James Lindemann Nelson (1989) Cutting motherhood in two: Some suspicions concerning surrogacy, and Kelly Oliver, Marxism and surrogacy, both in *Hypatia, A Journal of Feminist Philosophy* **4**, 85–115; R. Alta Charo (1988) Problems in commerialized surrogate mothering, in *Embryos, Ethics and Women's Rights*, (Elaine Baruch et al., eds.), Haworth Press, NY, pp. 195–201.

[42]e.g., Bopp and Burtchaell, *Report of the Human Fetal Tissue Transplantation Research Panel* I.

[43]e.g., Robertson, *Hastings Center Report* 18, and Andrews, *Hastings Center Report* 16.

[44]e.g., I have argued that even if abortion is wrong, the pregnant woman's consent is necessary so long as she is affected by the decision

([1987]*Hastings Center Report* **17**, 13); cf. Lisa Cahill (1988) *Report of the Human Fetal Tissue Transplantation Research Panel* **II**, D58–D65.
[45]Mahowald et al., *Hastings Center Report* **17**, 11–12.
[46]Center for Biomedical Ethics, University of Minnesota, 211–231.
[47]*Ibid.*, 212.
[48]*Ibid.*, 213.
[49]*Ibid.*
[50]*Ibid.* 215–216.
[51]Mahowald et al., *Science* **235**, 1308.
[52]Center for Biomedical Ethics, University of Minnesota, 211.
[53]*Ibid.*, 224.

The Moral Significance
of Brain Integration
in the Fetus

Thomas A. Shannon

Introduction

The moral significance of the human fetus continues to pose difficult ethical questions to the majority of the population. If one abstracts himself or herself from either the pro-life or the pro-choice extreme, one finds a difficult range of ethical dilemmas. On the one hand, as a living entity with the human genome, one must affirm value to the embryo; on the other hand, we need to take into account other values external to the fetus. Thus, the difficult ethical questions are how to value the fetus and what status has that value in comparison to other values.

Frequently, this discussion has been couched in terms of a consideration of whether or not the embryo or fetus is a person. One level of this discussion was developed in an article entitled "Reflections on the Moral Status of the Preembryo."[1] This article specifically asked whether or not the preembryo as a biological entity could bear the claim of personhood ascribed to it by some. The article concluded that because the preembryo was not yet an

From: *Biomedical Ethics Reviews • 1991*
Eds.: J. Humber & R. Almeder ©1991 The Humana Press Inc., Totowa, NJ

ontological individual, i.e., an individual with its own intrinsic principle of unity, it could not claim the values and level of protection ascribed to persons. However, because it was a living entity and a member of the next human generation, it possessed an ontic value. That is, as an entity with a teleological unity (i.e., the developing cells are on their way to becoming an ontological individual), it is living and has a distinct genotype that it will share with no other, absent multiple births.

The argument in that essay is that the preembryo has a developing moral standing as it moves into each new stage of its biological evolution. Thus, the fact that the preembryo is not an ontological individual is the basis for the assertion, on the one hand, that the preembryo has moral standing as the next human generation, but cannot make a claim for the stronger status of personhood. Therefore, the protection of the preembryo can be qualified by the value of a person whose claims may come into conflict with it.

The purpose of this essay is to continue this line of argumentation by moving to the next level of moral significance for the consideration of the issue of personhood: brain integration. The argument is twofold: a review of the relevant data of fetal brain development to ascertain various levels of integration, and a consideration of the moral implications of these data for the standing of the embryo and fetus.

The Maturation of the Fetal Brain

The process of fertilization initiates a most remarkable and stunning set of biological developments resulting typically in a fully formed human being. Even more interesting is the fact that the central nervous system is "the first system to begin and probably the last to complete development..."[2] In this paper, the focus is on neural development, and thus, the description will pick up with the development of the primitive streak at the beginning of the third week of gestation.

The first phase of neural development is called neurullation and extends from about the fifteenth to the thirtieth day of development. During this stage, the primitive streak thickens at the midline of the embryonic disk and gives rise to mesenchymal cells. These then "migrate widely and have the potential to proliferate and differentiate into diverse types of cells..."[3] Some of these cells migrate toward what will be the head and become, around the fourth week, the neural plate. The cells of the neural plate become the neural tube that then differentiates into the central nervous system, consisting of the brain and spinal cord, and the peripheral nervous system, consisting of various "cranial, spinal, and autonomic ganglia and nerves."[4]

Around days 19–21, there is a division of the neural tube into the forebrain, midbrain, and hindbrain, together with the presence of the spinal cord. This phase concludes with the closure of the neural tube around day 30. The second phase is called canalization and lasts from about day 30 to day 52.[5] During this time, the cerebral hemispheres become distinct entities and the cerebral cortex begins its development. The development of synapses occurs very early in this period, but the major period of synaptogenesis is "during the seventh month of gestation, and continues until about 18 months postnatal."[6] Overlapping with this period is phase three, "Retrogressive Differentiation,"[7] which lasts from about day 46 to birth. During this phase, there is a remodeling process of some early structures. This period represents the full developing and integration of all the structures of the central nervous system. Of particular importance during this period is myelination, during which various cells, as a kind of insulation and strengthening material, wrap themselves around the somatic motor neurons and the autonomic motor neurons.[8] This process

is essential for the normal development of the brain and spinal cord, and it may even be that an individual's acquisition of behavioral and psychological capacities is dependent on the extent of myelination of appropriated brain regions.[9]

Also during this period, we have the first establishment of a neural network in which the cerebral cortex becomes connected with various regions of the brain and the spinal cord. Such integration is critical because it incorporates the neocortex into the rest of the central nervous system.

Within the context of this structure of development, the developing embryo is beginning to present a variety of activities. Flower, for example, reports on the first reflex movements in a 7.5-week-old embryo. The neural basis for these movements is the result of a "simple, three-neuron circuit."[10] The transition from such simple reflex arcs to total body responses requires more neural circuitry. This is initiated, beginning in the ninth week, with the development of the brainstem reticular formation. Flower suggests that this system

> could produce a repeated output that was a modulated version of its intrinsic neural activity *and* sensory inputs received. This sort of system might then account for motor activity *in utero.*[11]

Of critical importance is the transition from whole body movement to specific local reflexes. Flower notes three characteristics of this more sophisticated neural activity:

1. Electrical activity becomes more regular after the tenth gestational week,
2. Some neuromuscular systems must be inhibited for other local reflexes to take place, and
3. An increase in the capacity to respond to external and internal stimulation.[12]

Thus, beginning with the eighth week, there is a significant increase in the number of synapses and the developing capacity for movement increases the sensory input to the brainstem.

Other developments are of significance. Whereas the neocortex begins its development during the eighth week, the first synapses there do not occur until between the nineteenth and twenty-second weeks. This is marked by the presence of spiny stellate cells:

> the synaptic targets of incoming as well as intracortical neural pathways; therefore, their presence marks neocortical readiness to establish functional circuitry.[13]

The connection of neural pathways to the neocortex through the thalamus, "a multi-component structure which modulates sensory input just before relaying it to the cerebrum,"[14] is a critical state of development. This connection integrates the nervous system and, should this not occur, the neocortex would remain isolated. This occurs somewhere around midgestation. Thus, by about 20 weeks of development, the fetus has an integrated neural system.

Electrical activity can be detected around 6.5 weeks and probably "during the first half of gestation originates in the brainstem."[15] The first patterned EEG can be obtained around 20 weeks, and after 30 weeks an EEG pattern distinguishing sleep and wake cycles can be detected. Flower indicates two transitional periods:

> a possible consolidation of brainstem influence over motor activity and sensory input near the end of the first trimester, and establishment of the sensory input channel to the neocortex via the thalamocortical connection around mid-gestation.[16]

This means that at the conclusion of the embryonic period at about eight weeks, the embryo has the capacity of limited responses to stimuli. Flower also suggests that the experience of pain[17] might not be present until midgestation when the thalamocortical connection is made since "the noxious stimulus pathways pass through the thalamus."[18] In between these two points of basic sensation and awareness is the recognition that the fetus has "a system-modulating brainstem (at 12–14 weeks)."[19]

In a complementary study, Sass distinguishes between two phases of what he calls brain life. Brain Life I extends from fertilization until the seventieth day. During this stage, we have the development of the neural plate, cerebral vesicles, and the cortical plate. Brain Life II is the period after the seventieth day to maturity, which is concluded after birth. Critical in this second stage are the beginnings of synapses, "the precondition for all further intercommunicating between neurons and the maturation of the functioning of the brain organ."[20] Also during this stage, various interconnections develop between various parts of the brain and allow it to interact with other organs.

In summary, then, several factors are critical. First, the nervous system begins to develop quite early in embryonic life. Then we have the process of neurulation, which sets the immediate stage for the development of the whole nervous system. Third, the first neuron circuits appear around eight weeks. This is followed by the development of the reticular formation through which signals can travel down the spinal cord. Fifth, mylenation, which begins around the fourth month, is critical in helping form the integration of the nerve tissues. Finally, around the twentieth week, the thalamus becomes connected to the cortex with the result that the nervous system is physically integrated. Clearly from this point on, much more development is necessary for neural maturation, but the essential system is in place by midgestation.

Moral Implications

Introduction

The direction the consideration of the moral significance of these biological data will take incorporates and builds on previous discussions of the concept of brain life. That is, the question is whether or not these biological data about various stages of neural development are also relevant for determining various levels of moral standing with respect to the embryo or fetus. Such discussions go back to at least 1972.[21] A primary motive for the use of such a term was to create a sense of balance or symmetry with the term brain death. That is, if the death of the brain as an integrated entity was morally relevant as a marker of the death of the person, why could not the birth of the brain have a similar moral relevance at the beginning of life? Thus, again we have an affirmation of the central role of the concept of personhood in this debate.

Criteria for a Brain Life Standard
A Review of Perspectives

The use of the concept of the death of the brain has given rise to the analogous use of the birth of the brain, or brain life, as a means of determining either the beginning of personhood or yet another morally relevant developmental stage in fetal development. This section will present five different approaches to perspective. Commentary on these perspectives will be presented in the conclusion.

JOHN M. GOLDENRING: Goldenring first proposed the concept of brain life in a dissertation entitled *Death Life and Abortion: The Implications of the Fetal EEG*. The essential argument is that the "fetus is biologically a human being at the point at which its brain begins to function."[22] This means that, once the fetal brain is functioning, "one cannot advance any logical argument to show that that fetus is not a living human being, at least from the point of view of medicine."[23]

For Goldenring, the critical point of such brain formation is

the eight-week point...since there is no doubt at that point that an active brain by electrical and anatomic definition is clearly present.[24]

Such a time period reflects the integration of the brain as a whole and provides a symmetry with the death of the brain as a whole. Thus, the emphasis is on the

central nature of the brain in defining a human being throughout life, and not the specific point at which we define the
presence of the operating brain.[25]

Goldenring recognizes that biological facts are not going to
resolve our ethical questions because the critical question is how
we value such facts. But nonetheless, he thinks that this orienta-
tion "can help clarify decisions to the point where areas of potential societal compromise can become clearer."[26]

HANS-MARTIN SASS: Sass proposes two critical stages in fetal
brain development. Brain Life I occurs around the fifty-fourth
day, post conception, which is characterized by "the appearance
of the cortical plate, formed by post-mitotic stationary cortical
neurons."[27] Although there are yet no synapses at this stage, what
is significant is that this stage "represents the first living cells,
some of which will, as a result of further development, function
as vital parts of the human cortex."[28] Brain Life II begins after the
seventieth day, post conception, and is characterized by the first
formation of synapses. Here "the post-mitotic cortical cells are
now beginning to interconnect and communicate in isolated areas
that will form ever growing networks of interconnections."[29]

Sass suggests using Brain life I as the point "after which fetal
life should be morally recognized and legally protected..."[30] This
stage is significant because it represents the earliest stage of tissue development, which is critical for the later development of
the full brain. Also, this perspective "matches the widely accepted criteria for brain death..."[31] Although these biological data
do not provide us with a moral value, they give us a criterion [a
basis] on which to base public policy without entering into "metaphysical or religious interpretations."[32]

THOMASINE KUSHNER: Kushner approaches the concept of
brain life from a psychological perspective rather than a biological one. She begins by distinguishing between *zoe,* which means
"*being alive* in a biological sense," and *bios,* in which individuals
"*have lives.*"[33] *Zoe,* the root for zoology, refers to life in the

physical sense, the organic structure and organic functioning of an organism. *Bios,* the root word for biography, refers to being "the subject of a certain life with its accompanying history, nexus of personal and social relationships..."[34]

The argument Kushner proposes is that it is "only a functioning brain [that] makes the consciousness possible on which bios depends."[35] Thus, the morally significant question with respect to the concept of brain life is whether or not the fetus is the subject of a life. Kushner's answer is that

> until it has developed a brain capable of consciousness the fetus' biography is not yet started. There is no life *(bios)* of which the fetus is the subject, although there are lives of which the fetus is a part.[36]

Kushner does not give a specific time when she thinks this level of functioning would occur. Quite obviousy though, this would be rather late in fetal development, for such functioning would assumedly require at least an integrated central nervous system or initial stages of neocortical development.

CAROL TAUER: Tauer distinguishes between two traditional concepts of the person, and then proposes a third perspective. First, we have the person strictly understood. These are the sorts of beings "who are moral agents, who have moral rights, and who are to be respected simply because they exist."[37] These are individuals who have rationality, can assume responsibility for their actions, and are self-conscious. Second, we have persons in the social sense. These are "all those whom our society recognizes as having the status of persons."[38] No specific criteria are associated with this definition because such criteria will, assumedly, vary from society to society.

Third is Tauer's proposal of the psychic person. Such a concept requires two conditions:

> (1) the present capacity to retain experiences as "memories" through the building of pathways in the central nervous

system; and (2) the potential to become a person in the strict sense.[39]

She argues that person in the psychic sense is morally relevant because of two linking factors:

> (1) that there is a continuity of experience between a person in the psychic sense and the person in the strict sense into which he or she develops; and (2) that the experience of a person in the psychic sense begins to determine the development of the personal psychological characteristics of that particular individual.[40]

Thus, the critical question for Tauer is how far back one's identity may extend. This is because "It is only I as an experiencing whole who can even begin to have moral value."[41] Thus, critical for such identity is the development of physiological structures to support such memories and to permit one's physiology to influence one's development. Tauer accepts

> any evidence of brain activity as an operational standard for psychic personhood, thus setting the initial time at approximately 6.5 weeks gestational age—or, more generally, during the seventh week of fetal development.[42]

D. GARETH JONES: Although not totally convinced that the idea of brain life is a useful one, primarily because of the difficulty of recognizing its presence,[43] Jones does recognize that there are some merits to pursuing the analogy with brain death. Accordingly, Jones says

> It is not unreasonable to argue that, at some particular stage during development, numerous discrete functions come together so that the rudimentary nervous system begins to function *as a nervous system.*[44]

The problem this creates for Jones is the fact that the critical factor of the developing brain is that "different systems are laid down at different times, and these are not coordinated until relatively late in development."[45] Thus, Jones argus that if the term brain life is to be used, it should be used "when most developmental sequences have started..."[46] This would be between 24–28 weeks.

The problem that Jones sees here is that if the brain birth concept is used to mark the transition from nonperson to person, then the date he sees as biologically relevant is quite late. On the other hand, if one were to use an earlier date, such as eight weeks, criteria "quite unlike those employed for brain death"[47] need to be used.

Conclusions: The Moral Relevance of the Concept of Brain Birth

Objections

Using two definitions of death by brain criteria as the bases for the analogy with brain life, Jocelyn Downie argues the analogy is not as good as one might think.

First, if we use the criteria of whole brain death, including the death of the brainstem, she argues that "no immediate connection can be drawn between brain death and brain life."[48] For example, this means that the definition of death is the cessation of the organism as a whole, and, correspondingly, brain life would be the "commencement of the functioning of the organism as a whole and it is possible that brain life is not the criterion for this commencement."[49] Additionally, if integration is the key, as another element paralleling the lack of integration in brain death, then it could be argued that "the embryo is functioning as an organism as a whole from the moment of fertilization."[50]

Second, if one uses the loss of that which makes us distinctively human as the definition, the criterion of which is the loss of neocortical function, other biological problems arise, according to Downie. First, it "violates the widely accepted requirement of transspecies applicability for a definition of death."[51] That is, since in other areas of the biological world, death is the cessation of the organism as a whole, why should human death be defined independently of the biology of that organism? Second, this definition excludes "from the category of the living those human beings that we commonly consider to be living."[52] The example she uses is a patient in a PVS state and suggests that this definition "describes the loss of something of essential significance rather that the death of a human being."[53]

Thus, Downie concludes that "The capacity to think/be conscious is not essential to the existence of a living human being."[54] The definition, therefore, cannot be applied to the embryo. Since a greater understanding about the role of the brain at the end of life does not necessarily yield conclusions about the brain at the beginning of life, especially with respect to the status of integration, the converse of the first brain death definition in not applicable. Finally, since the second definition of death is "flawed, any conclusions about brain life drawn from it should be viewed with suspicion."[55]

Another set of objections is raised by Jones, as previously noted. In general, his objections stem from the gradual development of the brain itself and the sequential, rather than simultaneous, way in which various capacities present themselves. Then too, Jones points to the difficulty of developing criteria or means by which such capacities can be measured or discerned. Thus, for example

> The EEG appears to be particularly unhelpful, since the electrical inactivity of brain death is a normal feature of the developing brain, albeit for periods of time lasting from a matter of seconds to a few minutes. This lasts until about 32 weeks gestation.[56]

Finally, Jones identifies a cluster of issues, primarily from a biological perspective, that show difficulties with the brain life concept.

We cannot get away from the major differences between a progressive phenomenon which is leading somewhere new, and a once and for all phenomenon which is the final point of an existence that is now at an end. In biological and clinical terms we are dealing with quite separate considerations, and it may be confusing to use the one as a model of the other. The alleged symmetry between the two is not as strong as sometimes assumed and has yet to be provided with a firm biological base. Further work on this issue will have to take account not only of the issues raised in this paper, but also of the contrast between the *order* of neural embryogenesis and the *disorder* of neural death, and therefore of the contrast between the healthy dimension of brain birth and the pathological dimensions of brain death.[57]

Justifications

Goldenring is the one who takes the analogy of brain death and brain life most seriously and bases this on the "central nature of the brain in defining a human being throughout life."[58] Thus, he can argue that "whenever a functioning human brain is present, a human being is alive."[59]

Also of critical importance for Goldenring is his position that, although recognizing that how we value facts and not the facts themselves resolve our social policy questions, "a soundly based scientific definition of a human being can clarify decisions to the point where areas of potential societal compromise can become clearer."[60] Thus, for Goldenring, the brain life definition, beginning for him at eight weeks, provides a criterion that is consistent, verifiable, and symmetrical between the beginning and end of life, and is based on relatively objective criteria.[61]

Sass set out a biological, philosophical, and theological frame of reference to justify the use of brain life. He begins by arguing that both the Graeco-Roman and the Judeo-Christian traditions "single out the capability of reasoning, communicating, and choosing values as the single criterion that sets humans apart from the rest of nature and gives them 'human dignity.'"[62] This suggests to Sass that the mere fact or presence of a functioning organism or biological existence does not confer legal or moral protection. Something transcending the biological is required.

Similar transcendent criteria on which to base the concept of brain life for our society would have to meet four requirements for Sass

> (1) compatibility with most of the major cultural traditions in existing pluralistic societies, (2) compatibility with the medical understanding and explanation of the process of human embryonal development and gestation, (3) simplicity of definition and diagnosis, and (4) a high preponderance of moral, medical, and cultural advantages over disadvantages.[63]

Sass is satisfied that his use of Brain Life I—54 days post conception—meets these criteria. Prior to this time, for example, "even the biological preconditions for being an *'imago Dei'* or *'zoon logon echon'* are not present."[64] Then too, as previously noted, Sass thinks the development of the cells of the cortical plate are morally significant because these are the ones that will develop into the cortex. Additionally, he thinks that such a criterion as Brain Life I can

> justify a consensus that will cut across a broad variety of historically developed moral and cultural positions in pluralistic societies. Its introduction would not force those who prefer to recognize earlier stages of human life such as gametes, fertilized or unfertilized eggs and morula, to abandon those beliefs, or to change their religious conduct.[65]

Sass also thinks that Brain Life I can be diagnosed reasonably well by counting post-ovulationary days, by ultrasound after diagnosing nidation, by measuring the seize (sic) of the embryo, or by following HCG testing.

Thus, Sass concludes that the biological criteria are meaningful and diagnosable for the concept of Brain Life I, and that such a concept is compatible with the Judeo-Christian and Graeco-Roman traditions. It is thus a reliable criterion for discerning embryonic life that is legally and ethically protectable.

Although Kushner does not ground her perspective on a clear biological criterion, she does highlight the moral difference between life understood as organic and as biography. The difference is between being a life and being the "*subject* of a certain life with its accompanying history, nexus of personal and social relationship, plus the whole fabric of events as they happen to and affect the individual."[66]

The critical point is not a devaluing of organic life, the reality of being a life, but rather the way in which being the subject of a life is morally relevant to differentiating the two. Thus, for human existence what is morally significant is the capability of biography, and the foundation of this is a brain developed to the point where it will support consciousness. Kushner also argues that persons "are not persons because they have a certain type of body."[67] That is to say, the morally relevant feature of personhood is the capacity of personal activity, the capacity for being a subject. This capacity is not conferred, in Kushner's perspective, merely from appearances. Again, the critical point is the distinguishing of biography from physical life.

Continuing the concept of the moral relevance of being, the subject of a biography is Tauer's presentation of the psychic person, the individual with the capacity to be a person in the strict sense and one with the capacity to retain memories and be shaped by one's environment. Such a capacity indicates the beginning of a continuous sense of identity, and this is grounded on the basis of a nervous system capable of sustaining such events. Tauer locates this at the seventh week, when fetal reflexes begin to appear.

Tauer recognizes that even though a person may not be present at this seven week mark, such a potential "may make serious moral demands on us."[68] Thus, her point is not to make the early fetus morally irrelevant, but rather to indicate the moral difference between stages of development and to indicate that different values and arguments for the protection of such an entity are needed that specifically correspond to that level of development.

Discussion

What this review of the biological and philosophical/theological literature has shown is a very complicated and marvelous picture of the development of the embryo and fetus. One cannot study any of this literature without a sense of wonder or awe at this almost incredible process of development that leads to the adult human.

Difficulties with Brain Life

Yet, as amazing and wondrous as this reality is, several elements emerge that are critical in assessing the status of the embryo and fetus. First, a general problem from this perspective is the danger of identifying or locating the self within the brain. That is to say, given the perspectives of Sass and Goldenring in particular, the brain takes on a role that is central in defining a human being and, consequently, the self. This may redirect us to a revival of Cartesianism, which seems to locate the self within or confine it to the brain. Additionally, such a perspective presents an exceptionally narrow concept of the self by totally neglecting the body and the social context in which one lives as key dimensions of self-identity. Thus, the concept of the human or the self based exclusively on the brain perspective provides a very narrow and solely biological understanding of the self.

Second, such a perspective may be reductionist in that the position is open to an identification of the self with the brain. Nothing more is needed for the self than the functioning brain— at whatever level one determines is appropriate. Of course, such a determination is based on physiological criteria that are self-referencing. Thus, the circle is completed so that one can argue that the self is the brain.

Third, one needs to distinguish between biography and *auto*biography. Whereas Kushner's point is well taken, we must remember that a biography can be written by anyone. What is distinctive about autobiography is that it is the record of the self by the self, and this typically includes more than reporting mental events. Also, assuming that the subject of the autobiography arises with the functioning brain begs the question of the relation of the self to the brain. It also presumes a certain understanding of the self.

Benefits of Brain Life

Nonetheless, given these critiques, I think there is merit in the concept of brain life. First, and I think most critical, is that the organism passes through several stages of development. There are relatively clear markers for new phases of the growth of the embryo and fetus. Although Jones' point of the gradualness of some of these stages must be taken seriously, nonetheless, there are markers of development and these can be identified with reasonable accuracy. At each stage of development, a new capacity comes forth, or one that is present is further enhanced. A gradual maturing and integration of various emerging systems is characteristic of fetal growth.

Second, because of this process of growth, all stages are morally differentiated. That is, although it is obvious that if there is no overall growth, there will be no organism, nonetheless, some processes serve as the precondition or ground of a particular characteristic that sets the stage for the emergence of a different capacity. Thus, the early presence of synapses makes possible reflex activity, and this becomes the ground of experience. The development of the cortex and its integration with the spinal cord initiate the foundation of the sense of identity. That is, various levels of biological development are a necessary, but not sufficient, condition of personhood. The emergence of such markers takes on moral significance because of their relevance to the presence of a person.

Two distinctions are important here. One is the differentiation between organic life and personal life. This difference is important because the person's life is morally more significant than organic life and is entitled to greater levels of protection. This does not mean that organic life is not valuable; as life, it certainly is valuable as a premoral good. However, organic life is not the subject of an autobiography, it is not an end in itself, it does not have the capacity of freedom.

Second is the question of "whether the capacity of *a* to develop into *A* makes *a* the same as *A*."[69] The issue is whether there is a moral equivalence between the zygote, which, all things being equal, has the potential to become an adult human, and the seven month old fetus, which also has this same capacity, but is obviously at quite a different developmental stage. Clearly, the living and developing zygote has potential and is deserving of respect. Such respect, however, follows "from its potential, and not because its potential has been converted into an actuality."[70] Thus, one must recognize the moral implications of the different stages of development, and what is particularly of moral significance is the narrowing of the gap between potentiality and actuality as development occurs.

Summary

I conclude that the concept of brain life is of moral relevance insofar as it reveals another critical stage of development and grounds the biological presupposition for another capacity identified with personal life. I also concur in the differentiations of the capacities between the brain at eight weeks and twenty weeks. With several of the authors discussed, I recognize the moral significance of the development of the synapses and first neural connections. This first presence of neural activity marks the earliest time at which one could say the foundations of autobiography or personal identity begins. Yet it is also important to recognize the critical jump that emerges with the integration of the whole nervous system as a system, particularly the development and integration of the cortical areas of the brain. Such a biological reality, although not identical with the person, grounds the distinctive personal reality of self-consciousness in a truly significant way.

Thus, while the preembryo is both deserving and entitled to our respect because it is a living entity with the human genome, it is not morally equivalent to the actual human person because it is neither individuated nor has the beginnings of the neural system. This would suggest that, should there be a conflict of values between the preembryo and another critical value, that value could be chosen over the preembryo.

The eight-week-old embryo manifests early stages of neural activity and grounds the earliest possibility of personal identity. The twenty-week-old fetus presents an integrated neural system that grounds the possibility of self-consciousness. In these latter two cases, the gap between potentiality and actuality is closing, and as such, these entities are entitled to greater degrees of respect and protection. Although they are not yet morally identical to actual persons, this actuality is more firmly grounded in their rapidly maturing physiologies, which, in turn, enables the next level of maturity to emerge. Whereas none of these entities is entitled to the same degree of respect or protection accorded to actual persons because of the absence of biological preconditions necessary (but not sufficient) for personal activity, their own levels of development ground a degree of respect and protection in proportion to their establishing the biological presupposition for morally relevant personal capacities.[71]

Thus, should there be a conflict of values between the embryo prior to Brain Life I and another individual past that stage or some other critical value, such as a duty to care for another or, in an extreme situation, the triaging of medical care, a decision in favor of the other person or other value could be made. That is, even though individuality has been established, the capacity for autobiography has not. Because of the lack of the biological presupposition necessary for autobiography, other persons or critical values could be given higher priority. Such conflicts must be significant and the values at stake serious. This means that the realities of individuality and possession of the human genome, when combined with Brain Life I, have a developing but not ultimate claim for moral and legal standing within the human community. Because such claims are not total or full, other persons or values can be given priority.

Notes and References

[1]Thomas A. Shannon and Allan B. Wolter, OFM (1990) Reflections on the moral status of the preembryo. *Theological Studies* **51**.

[2]Ronald J. Lemire and Josef Warkany (1986) Normal embryology, in *Disorders of The Developing Nervous System: Diagnosis and Treatment* (Harold J. Hoffman and Fred Epstein, eds.), Blackwell Scientific Publications, Boston, MA, p. 3.

[3]Keith L. Moore (1988) *The Developing Human: Clinically Orientated Embryology,* 4th ed., W.B. Saunders Co., Philadelphia, p. 55.

[4]Moore, *The Developing Human,* p. 364.

[5]Lemire and Warkany, Normal embryology, p. 7.

[6]D. Gareth Jones (1989) Brain birth and personal identity. *The Journal of Medical Ethics* **15,** 177.

[7]Lemire and Warkany, Normal embryology, p. 7.

[8]Moore, *The Developing Human,* pp. 373–74.

[9]Jones, Brain birth and personal identity, p. 176.

[10]Michael Flower (1985) Neuromaturation of the human fetus, *The Journal of Medicine and Philosophy* **10,** 239.

[11]Flower, Neuromaturation, p. 240. Italics in the original.

[12]Flower, Neuromaturation, pp. 241–242.

[13]Flower, Neuromaturation, p. 243.

[14]Flower, Neuromaturation, p. 243

[15]Flower, Neuromaturation, p. 245.

[16]Flower, Neuromaturation, p. 246.

[17]Pain and suffering are not to be identified. Pain is a biological experience. Suffering is an interpretation of the meaning or significance of that pain. Neither necessarily implies the other, but a certain level of self-consciousness needs to be present to perceive suffering.

[18]Flower, Neuromaturation, p. 247.

[19]Flower, Neuromaturation, p. 247.

[20]Hans-Martin Sass (1989) Brain life and brain death, *The Journal of Medicine and Philosophy* **14**, 51.

[21]J. M. Goldenring (1982) The development of the fetal brain. *The New England Journal of Medicine* **307**, 564.

[22]John M. Goldenring (1985) The brain-life theory: Towards a consistent biological definition of humanness. *The Journal of Medical Ethics* **11**, 198. Italics in original.

[23]Goldenring, The brain-life theory, p. 199.

[24]Goldenring, The brain-life theory, p. 200.

[25]Goldenring, The brain-life theory, p. 200.

[26]Goldenring, The brain-life theory, p. 204.

[27]Hans-Martin Sass (1989) Brain life and brain death: A proposal for a normative agreement. *The Journal of Medicine and Philosophy* **14**, 51.

[28]Sass, Brain life, p. 52.

[29]Sass, Brain life, p. 52.

[30]Sass, Brain life, p. 52.

[31]Sass, Brain life, p. 57.

[32]Sass, Brain life, p. 58.

[33]Thomasine Kushner (1984) Having a life versus being alive. *The Journal of Medical Ethics* 10 (1984): 6. Italics in Original.

[34]Kushner, Having a life, p. 6. Italics in Original.

[35]Kushner, Having a life, p. 6.

[36]Kushner, Having a life, p. 6.

[37]Carol A. Tauer (1985) Personhood and human embryos and fetuses. *The Journal of Medicine and Philosophy* **10**, 255.

[38]Tauer, Personhood, p. 255.

[39]Tauer, Personhood, p. 259.

[40]Tauer, Personhood, p. 259.

[41]Tauer, Personhood, p. 261.

[42]Tauer, Personhood, p. 263.

[43]D. Gareth Jones (1987) *Manufacturing Humans: The Challenge of the New Reproductive Technologies.* InterVaristy Press, Leicester, England, p. 119.

[44]Jones, *Manufacturing Humans,* p. 119. Italics in Original.

[45]D. Gareth Jones (1989) Brain birth and personal identity. *The Journal of Medical Ethics,* **15**, 177.

[46]Jones, Brain birth, p. 177.

[47]Jones, *Manufacturing Humans,* p. 124.

[48]Jocelyn Downie (1990) Brain death and brain life: Rethinking the connection. *Bioethics* **4,** 225.

[49]Downie, Brain death, p. 219.

[50]Downie, Brain death, p. 219.

[51]Downie, Brain death, p. 223.

[52]Downie, Brain death, p. 223.

[53]Downie, Brain death, p. 223.

[54]Downie, Brain death, p. 225.

[55]Downie, Brain death, p. 225.

[56]Jones, *Manufacturing Humans,* p. 122.

[57]Jones, Brain birth, p. 178.

[58]Goldenring, The brain-life theory, p. 200.

[59]Goldenring, The brain-life theory, p. 200.

[60]Goldenring, The brain-life theory, p. 204.

[61]Goldenring, The brain-life theory, p. 202.

[62]Sass, Brain life, p. 48.

[63]Sass, Brain life, p. 50.

[64]Sass, Brain life, p. 52.

[65]Sass, Brain life, p. 56.

[66]Kushner, Having a life, p. 6.

[67]Kushner, Having a life, p. 7.

[68]Tauer, Personhood, p. 263.

[69]Jones, *Manufacturing Humans,* p. 146. Italics in original.

[70]Jones, *Manufacturing Humans,* p. 147.

[71]I would like to thank my colleagues Mario Moussa and Ruth Smith for their reading of the manuscript and providing very helpful comments.

The Embryo as Patient

New Techniques, New Dilemmas

Andrea L. Bonnicksen

"Conception (in contrast to the fully public fact of birth) suggests not only the unknowable but the forbidden: our birth dates are matters of public record but our dates of conception are permanently shrouded in mystery."[1]

In the few years since Joyce Carol Oates described conception as "shrouded in mystery," advances in reproductive technologies have further eroded the aura of the unknown in conception. Human eggs are fertilized in the laboratory, stored as embryos in glass straws in freezing tanks, and subjected to various therapeutic and nontherapeutic manipulations. As studies on embryos increase in frequency and variety, and as they move from the laboratory to the physician's office, they give new urgency to the need to promote an ethical debate about the benefits of embryo micromanipulation for couples, embryos, and society at large. This chapter reviews emerging techniques in embryo micromanipulation and identifies ethical issues that must be addressed before the techniques are systematically offered in medical clinics.

From: *Biomedical Ethics Reviews • 1991*
Eds.: J. Humber & R. Almeder ©1991 The Humana Press Inc., Totowa, NJ

Overview

Practitioners of in vitro fertilization (IVF) constantly try to improve IVF's success rate by varying such things as the culture media and the number of cells reached before the embryos are transferred. Embryo freezing, now a part of most IVF programs, has introduced still further refinements, such as the speed and method of the freeze and thaw. The embryo, in short, has long been a research subject in IVF.[2] This paper deals with something slightly different—the embryo as "patient" subject to diagnosis and/or treatment. Techniques for diagnosing and treating embryos are still experimental or only envisioned. The momentum is building quietly and quickly for embryo manipulations in the clinical setting, however. Although the difficulties of working with tissue as delicate and complex as the human embryo will slow the inclinations of practitioners to move ahead, several events suggest that new reproductive techniques and new dilemmas are imminent.

A growing number of published studies attests to active embryo research among teams in the US and abroad. Other countries, notably Britain, have set up procedures for reviewing embryo research.[3] Rapid advances in animal embryology suggest the range of techniques and the body of knowledge that can be extended to human embryos. Scientists and practitioners are speaking with an increasingly persistent voice about the need for embryo studies,[4] and published reports often conclude with observations about the implications of findings for clinical application. Thus, despite a generally inhospitable political climate for embryo manipulations,[5] research is quietly proceeding that is underwritten by the host institutions or by corporate grants. These published studies have fed a momentum for applications in which the embryo is the "patient." This field results from the confluence of rapidly developing specialties in IVF, cytogenetics, embryo micromanipulation, and molecular biology.[6]

Clinical embryo manipulation will raise ethical questions mirroring those in all areas of medicine, such as the need for informed consent, truth-telling, and confidentiality. Other ethical concerns have not directly been confronted, however, and the embryo as patient demands inquiry for the new dilemmas it raises. Before turning to those questions, the techniques of embryo biopsy, embryo microsurgery, and genetic therapy are reviewed.

Definition and Techniques

Definition of Embryo

Fertilization is a process, not a set moment.[7] During IVF, physicians mix eggs and sperm in a glass dish. If all goes well, the egg is fertilized in the first day and becomes a single-cell "zygote."[8] The zygote divides into cells and becomes a "preimplantation embryo." The embryo, still not visible to the naked eye, is transferred to the woman's uterus when it reaches four or eight cells. The rate of implantation (attachment of the embryo to the uterine wall) varies from center to center, with established centers reporting a 15–20 percent pregnancy rate.[9] Around the fourteenth day, the so-called primitive streak develops. This marks the time that "one is guaranteed that a single biologic individual is in the process of formation."[10] Neural and heart development later follows, and the embryo takes on the appearance and traits of a fetus.

For the purpose of this chapter, the "preimplantation embryo" or "embryo" refers to the tissues that cleave in the laboratory before transfer to the woman's uterus. Embryos can be kept cleaving outside the body for more than several days, but there is generally no clinical need for doing this. An embryo that cleaves to the point that it becomes a bundle of cells is known as a "blastocyst." Ethics commissions have agreed that the development of the primitive streak around day 14 is an important physiological development.[11] A 14-day embryo still outside the body (this has not been documented) would take on a "special moral status."[12] In 1979, the federally commissioned Ethics Advisory Board (EAB) concluded that the embryo is a "potential" human being "entitled to profound respect; but this respect does not necessarily encompass the full legal and moral rights attributed to persons."[13] Differing perspectives of the embryo's nature arise in public discourse,[14] but the EAB perspective is adopted in this chapter.

Embryo Biopsy

In the embryo biopsy, specialists remove one or two cells from the cleaving embryo to study for chromosomal or genetic defects.[15] If the biopsied portion is normal, the "parent" embryo, kept in culture or frozen, is transferred to the uterus of the woman undergoing IVF. If the biopsy reveals abnormalities, the parent embryo is discarded.

Eventually, biopsies will yield information similar to that available from the prenatal tests of amniocentesis and chorionic villi sampling. The difference lies in the timing. Embryo biopsies occur before implantation and during IVF; amniocentesis occurs after implantation. Likely candidates for embryo biopsy are couples trying IVF who have a history of miscarriages (sometimes caused by chromosomal abnormalities), or those at risk for passing a genetic disease to their offspring. An embryo found to be defective by biopsy will not be transferred to the woman. A fetus found to be defective by amniocentesis or chorionic villi sampling will be aborted or the test results will give a couple unwilling to terminate the pregnancy time to prepare for bearing a child with a handicap.

Researchers examining human embryos for chromosomal content have found a rather high rate of defects.[16] Chromosomal defects are responsible for Down Syndrome, Turner Syndrome, Klinefelter Syndrome, and other syndromes associated with mental retardation and/or physical abnormalities.[17] Most chromosomally abnormal embryos and fetuses are silently sloughed from the mother's body as a natural way of reducing birth defects.[18] Chromosomal abnormalities in embryos created during IVF, then, indicate the embryo will probably not implant after transfer.[19]

Tests to detect genetic abnormalities are more difficult and are less developed than tests to detect chromosomal defects. Researchers can extract a portion of DNA from an embryo's cell, amplify the region containing the target gene, and use a probe to detect the presence of a faulty gene.[20] Investigators already can use DNA amplification to determine the sex of embryos.[21] Rapidly unfolding genetic knowledge suggests that biopsies for genetic diagnosis for disease status are now "theoretically possible."[22] Several studies have pointed the way to detecting in embryos genes associated with the single-gene defects of cystic fibrosis and Duchenne muscular dystrophy.[23]

The embryo biopsy involves reducing the cell mass of the embryo. Is this a safe procedure? British researchers, under the watch of an agency that oversees research involving human embryos, studied the safety of the embryo biopsy by removing one or two cells from eight-celled embryos and watching the development of the parent embryo.[24] Most of the parent embryos developed to the blastocyst stage, leading the researchers to conclude that the biopsy was safe, although many variables still needed to be refined.[25] They expressed the opinion that clinical application was appropriate as long as the patients were informed of the risk and agreed to follow-up prenatal tests.[26] If the biopsy is shown to be safe, a logical extension of the technique will be to "twin" an embryo, as is already done in animal husbandry, and create a "duplicate" embryo to be preserved and transferred to the woman in the event the first embryo fails to implant.[27]

Embryo Microsurgery

Embryo microsurgery manipulates an embryo to correct an abnormal condition. It requires new skills, procedures, and equipment, and will likely have clinical applications by the end of the decade. Sperm microinjection illustrates what is now possible with microsurgery on reproductive tissues. If a couple cannot conceive because the man's sperm do not penetrate the woman's eggs, technicians, during microinjection, hold the egg still, drill an opening in the *zona pellucida* ("envelope" surrounding the egg), and insert a spermatozoan with a pipet.[28] At least one child has been born after having been fertilized by the manual insertion of a spermatozoan.[29] British researchers studying the safety of microinjection have concluded that the procedure is not correlated with chromosomal defects in the fertilized eggs. They expressed the opinion that the procedure was not correlated with defects and that this finding should "provide reassurance and support for the clinical implementation...of this method of dealing with male infertility."[30]

Sperm microinjection is a technology creating the need for more technology. Under traditional IVF procedures, approximately 5% of the time, more than one spermatozoan penetrates the egg, leading to an abnormal "polyspermic" fertilized egg with three or more (rather than the normal two) pronuclei.[31] Polyspermy is the cause of up to 20% of spontaneously aborted fetuses, and polyspermic zygotes are not transferred during IVF.[32] One side effect of sperm microinjection has been a greater proportion of polyspermic zygotes.[33] Thus, "epronucleation" is a new microsurgical procedure to remove the extra pronuclei. It has been attempted both successfully and with mixed results.[34] Researchers call for more studies with animal embryos before offering epronucleation in the clinical setting, either to correct natural polyspermy or to fix polyspermy caused by microinjection.[35]

Another microsurgical procedure is "assisted hatching." Unlike the procedures discussed above, this was performed on embryos transferred to women's uteruses, with the consent of the women, as therapeutic research. The investigators hypothesized that failure of the embryos to hatch from the *zona* is one cause of low implantation rates in IVF. To test this hypothesis, they made incisions in the *zona* of 115 early embryos before transfer to the women's uteruses. All the embryos were intact after the procedure. The procedure doubled the chances of implantation from the usual 11% to 23%, although it had the undesirable side effect of increasing the rate of multiple pregnancies.[36]

Gene Therapy

Gene therapy on human embryos has been anticipated with forboding and uncertainty. It has not been attempted in human embryos, but it is used for commercial purposes in animals. Gene therapy falls into two main categories: somatic cell and germ cell. Somatic cell therapy manipulates the genetic composition of a portion of an individual's body cells. The genetic alteration affects the patient only and is not passed from one generation to the next. It does not raise serious ethical questions beyond the need for rigorous safety testing and fully informed consent. The first experimental protocols involving somatic cell therapy on humans are now in process.[37] Somatic cell therapy on human *embryos* is still a distant possibility, however. If a late stage embryo has a genetic defect that prevents cells from producing a certain hormone, for example, the abnormal cells could be removed and replaced with corrected stem cells.[38]

With germ cell therapy, changes are introduced into the embryo's genes in a way that will direct the entire genetic structure of the potential individual and be passed to at least one generation. These manipulations raise serious ethical questions in that the genetic material of future generations is at issue.

Germ line manipulations are performed in mice, fish, goats, pigs, and other animals for commercial purposes. The rat growth hormone gene has been inserted into mouse embryo pronuclei, for example, to yield larger than normal mice.[39] At least some of the large mice, in turn, gave birth to large offspring. A research animal, Oncomouse, has been genetically engineered to produce offspring that rapidly develop tumors.[40] Pigs that produce large litters, slim pigs, sheep that produce manipulated milk for pharmaceutical use, and salmon that produce an antifreeze protein are among the germ line manipulations performed in animals for commercial reasons and passed on continually (a stable insert) or for several generations (an unstable insert).[41]

If germ line manipulations are advancing in animals, it is not unreasonable to expect them eventually to be attempted in humans, either to fix a defect, enhance a trait, or increase resistance to infections or disease. Observers have long worried about the possibility of enhancing socially desirable traits, such as height or intelligence.[42] Some scientists would draw the line at "enhancement genetic engineering,"[43] but the distinction between corrective and enhancement genetic alterations is unclear. At present, the National Institutes for Health (NIH) guidelines forbid human germ line research.[44] These guidelines affect recipients of federal grants, but do not apply to research done in private companies.

Ethical Issues

Arguments For and Against Embryo Manipulation

Proponents of embryo diagnosis and therapy argue that these techniques are ethically appropriate.[45] Diagnosing defects in embryos is preferable to diagnosing defects in fetuses in that it is less ethically problematic—and easier on couples—to discard a preimplantation embryo than to abort a fetus. Moreover, embryo diagnosis and treatment is a way of saving embryos that would have otherwise been discarded. Embryo therapy, in which defects are eliminated at the germ line stage and then passed on to the next generation, will help phase certain inherited defects from the gene pool. Embryo therapy to prevent the passing of disease is likened to vaccinations; just as we are morally obligated to vaccinate children to prevent the spread of disease across the current generation, so are we morally obligated to take steps to avoid passing a genetic defect across generations if the technology exists to do so.[46]

Supporters of embryo manipulation accept research in which embryos are studied and discarded. The loss of some embryos in research is not a problem provided guidelines are followed and the justification is strong. Many embryos are lost in nature in any event; embryos are potential life, not actual life; and research on embryos will lead to knowledge that will aid conception and, hence, the creation of life. The ethical cost of discarding embryos does not outweigh the expected benefit of research designed to improve conception rates, allow genetically at-risk couples to have children, and contribute to knowledge that will promote human health.

Supporters, in short, tend to see embryo diagnosis and therapy as benign efforts to prevent birth defects and to open a window to ways of resolving pressing problems. They have an optimistic view of reproductive technology. They see the benefits as concrete and the costs as largely symbolic; i.e., the embryo is a valued symbol of potential life. Steps can be taken to protect the symbolic value of the embryo while pursuing research benefits; limits can be imposed on inquiry as the need arises.[47]

Critics conclude that embryo diagnosis and therapy are ethically unacceptable for a variety of reasons.[48] Some argue that the embryo is more than potential life; it is life. As life, it has constitutional rights that include the right to life. Others focus on the research itself. The embryo cannot give informed consent for research and no one can give consent on behalf of the embryo. The research is degrading to and exploitative of women, who must be superovulated to yield enough eggs for fertilization for research purposes. Conducting research in which embryos are divided into experimental and control groups for systematic manipulations is calculating, dehumanizing, and harmful to societal values. Other arguments revolve around the goals of embryo diagnosis and therapy. Genetic diversity has evolutionary value; tinkering with it may have unfortunate consequences for future generations. Embryo therapy subdivides reproduction into smaller units that give physicians progressively greater control over what was once an autonomous and continuous reproductive process enjoyed by women. Embryo therapy wrongly places a value on perfection; children are lovable with or without defects. Embryo therapy will not stop at fixing defects in embryos for a small number of people; eventually the urge to breed desirable traits will win out and this will create new and troublesome power divisions in society.

Some critics, in short, tend to see embryo therapy as an assault on embryos to further a technological luxury that will open the door to medical and political control. Reproduction is a process better left to nature. Human involvement in the genetic makeup of humans is fraught with danger. The costs are concrete and the benefits speculative. The latter are not sufficient to justify the costs to women, embryos, potential children, and genetic diversity itself.

Debate over embryo manipulations has been ongoing since the earliest efforts to fertilize eggs outside the body. Until recently, the debate has been largely theoretical and conducted with a lulled feeling that the presence of restrictive laws in some states, the potency of the right-to-life movement, and the absence of federal funding for embryo research will make clinical applications with embryos an elusive development. The push for research and publication of studies demonstrate that we are on the brink of clinical applications, however. It has been said that the question of embryo research is intractable.[49] Although the moral rightness of discarding embryos may never be settled, the strength of that debate must not turn our attention from the events that are taking place and that need addressing now. Below, four questions about the embryo as patient are raised that should be addressed before further clinical applications are contemplated.

Why Are We Doing This?

Embryo manipulation opens a new field of medicine in that a new entity—the embryo—becomes the subject of testing and treatment. One must ask, however, whether this emerging field meets a compelling need. It will save some anguish for genetically at-risk couples who might otherwise abort a fetus shown by prenatal testing to inherit the troublesome genetic trait. It might also increase pregnancies in IVF programs. There are, however, less dramatic alternatives for helping couples procreate. The genetically at-risk couples could try IVF with donated eggs, sperm, or embryos; turn to adoption; try prenatal screening (the odds are generally in favor of a healthy fetus); or forego reproduction. These are imperfect alternatives, but they may pose fewer problems for the at-risk couple than embryo testing and treatment. Is it really in the interests of the at-risk couples to turn to IVF in order to have laboratory access to their embryos? In vitro fertilization is an expensive, emotionally difficult procedure with a stubbornly low success rate. Why should a new group of people (those genetically at risk) turn to medically assisted conception? Does this *meet* a compelling need or *create* a perceived need?

Embryo manipulation is justified as a way of improving IVF's success rate and resolving an increasingly broad array of infertility problems. Again, one wonders whether this need is sufficiently compelling to justify an emerging field of medicine. Sperm microinjection, for example, combats male infertility, but the procedure leads to an increased chance of polyspermy. Thus, the technique of epronucleation rids the embryo of its extra pronuclei introduced by the technique of microinjection. One technique introduces a problem to be resolved by another technique. A less burdensome alternative is to avoid all this by using a sperm donor. It is understandable that the husband wants to be the genetic father of the child. Is that wish so important as to justify the increased burdens the new procedure poses to society and to the couple?

In a simpler alternative, the couple could use sperm donors at a clinic for less than $1000. The woman would go to the clinic each month and have the sperm inserted in seconds. Using sperm microinjection, in contrast, the couple must go through IVF. The woman is hyperstimulated, her eggs are extracted, a specialist microinjects the eggs, and epronucleation is performed if more than one spermatozoan penetrates the egg. The couple may also have the embryos biopsied. If spare embryos result, they can freeze them for later transfer. Is it really in the couple's best interest to do all this? Does the wish for a genetic child justify these personal and societal resources? Embryo testing and therapy introduce a "new line" of reproductive choices for consumption. Before that line is marketed and tested it is important to consider who it benefits and why it is being introduced. Are we opening this field of medicine to meet needs that cannot be resolved in any other way? Or do researchers gain intellectual advantage, practitioners gain new clienteles, and pharmaceutical companies gain new markets?

The "black box" is the unknown between what happens between conception and pregnancy.[50] Its presence frustrates scientists who yearn to have answers to the mysteries of conception. Are embryo studies pursued to meet clear and present clinical needs, or are they pushed to satisfy the curiosity of researchers with generalized ideas about the value of embryo knowledge for understanding disease some time in the future? Among other things, we need a rigorous assessment of who will benefit from embryo manipulation to understand the reasons we are inviting a new medical "patient."

Will the Costs Be Distributed Equitably?

Embryo manipulation does not directly use public funds in that the federal government does not fund projects involving human embryos nor do insurance programs for public employees cover IVF costs to patients. The costs of embryo studies are absorbed by research institutions, financed through contributions by couples to IVF clinics, or funded by corporate grants.

Embryo research will indirectly absorb public resources, however. Public money funds animal embryo research, which yields knowledge that can translate to understanding of human embryos. Animal embryology occupies the attention and time of quality researchers. Human embryo research funded privately will also attract the attention of specialists in the growing OB/GYN subfield of infertility. Might physicians and scientists look at other questions if they were not concerned with embryology? What would these questions be?

Embryo manipulations will also indirectly take resources because of the expense of setting up clinics for the new procedures. The equipment is specialized and expensive. A fully operating clinic will need tools for IVF, micromanipulation, and DNA analysis.[51] If the clinic is set up in a hospital, what types of health care will be overlooked when funding goes to the clinic (assuming a steady sized pie of allocations)? Will embryo manipulations divert the skills of experts and the funds from agencies from other fields? If embryo manipulation is elevated as a medical priority, what goals drop as a consequence on the hierarchy of values?

Embryo manipulation can remain a specialized procedure for a small clientele or become a choice for an increasingly larger clientele. The most restrictive use would be to offer it only to infertile couples trying IVF who are at genetic risk for passing on a defect to a child—for example, a woman with blocked fallopian tubes who is a genetic carrier for cystic fibrosis and whose husband is also a carrier. A less restrictive use would be to offer it as a routine standard of care for all couples trying IVF, even if they are not genetically at risk. One could open the options even more by offering it to fertile couples who are genetically at risk, but will not, for personal reasons, terminate a pregnancy (hence, making prenatal diagnosis unhelpful) or to fertile couples not at risk, but also unwilling to terminate a pregnancy.[52]

The broader the clientele, the more will resources be diverted to equipping and staffing centers with embryo diagnosis and treatment. Embryo manipulations are an ethical issue if they create a demand for sophisticated techniques that divert resources and attention from social and environmental causes of birth defects and child illness. Pursuing specialized techniques to promote child health when ill-health caused by social conditions stubbornly persists raises serious questions relating to distributive justice.

Another question relates to publicly funded access of couples to the techniques of embryo diagnosis and treatment. Embryo manipulation is experimental, in the eyes of insurers, as are IVF and embryo freezing. As a result, IVF is covered only in part by only some insurance companies, although five states mandate limited insurance coverage.[53] Embryo manipulations, then, will probably be available only to those who can afford it or who have exceptional insurance policies. This denies equality of opportunity for all couples to use the techniques. If embryo manipulation expands in use, important questions of fairness and equity will be raised in that poorer people will effectively be barred from using reproductive techniques available to wealthy parents to promote child health.

What Emotional Challenges Will Couples Face?

Embryo manipulation is presumed to be a benefit to couples, even though risk attends the procedures. If couples are advised of the obvious costs of financial, safety, and incomplete risk data, they still may face unexpected emotional challenges. Like couples trying IVF and embryo freezing, they will be thrust into uncharted psychological terrain. Couples develop attachments to their embryos and it is logical to expect an emotional reaction to embryo manipulation.[54] Suppose a couple has four embryos and three of them have chromosomal defects and cannot be transferred. In regular IVF, all embryos would be transferred and the three defective ones perhaps sloughed from the body with no one the wiser. With embryo biopsies, however, the couples will know they produced three defective embryos. Will this have an impact on the wife and husband's self-images? Will it introduce new worries that did not exist before? Will it put them into a new round of experimental techniques that will yield more disappointment than anything else?

Assume, in a futurist scenario, that a predisposition to colon cancer is detected through preimplantation diagnosis. How should the couple be counseled about transferring the embryo to the wife's uterus? Should the couple go ahead with the transfer or not? What if the gene for a moderate defect (e.g., cleft palate) is detected? What are the couple's alternatives? As one observer asks, "Who will help the couple to decide which embryos to reimplant, given a perplexing number of trade-offs in relation to likely susceptibilities to adult diseases? What criteria should we use to make these decisions?"[55]

The dilemmas of preimplantation diagnosis are in some ways less intense than those presented by prenatal diagnosis in that it is arguably easier to discard an embryo than to terminate a pregnancy. Yet, it would be a fallacy to suggest that no emotional dilemmas arise. In both cases, couples operate in the context of statistical uncertainty and take responsibility for a decision affecting their child's health. With embryo manipulation, the couple may, after years of infertility, have a deep emotional and financial investment in embryo transfer. To discard embryos is no small thing for infertile couples trying IVF. Moreover, if the couple opts to treat the defective embryo (a future possibility), the treatment may affect not just their child but their grandchildren as well. They are taking responsibility for the health of their cross-generational progeny.

Embryos have emotional meaning for couples. The more things are "done to" embryos, however, the more will physicians and technicians see them as objects to be studied, manipulated, and fixed. Already, two-day embryos are graded on a five-point scale for their transfer quality on the basis of the evenness of their blastomeres, presence of cellular debris, and fragmentation. Grade 1 is the highest evaluation (no fragmentation, even blastomeres) and Grade 5 is the lowest ("totally degenerate").[56] Embryo manipulation will open new ways of evaluating and making an object of the embryo. In the meantime, the embryo is a precious symbol of potential parenthood to the couple. Embryo manipulation demands clinical models that take into account the peculiar nature of the relationships among the couple, embryo, and physician.

Where Should We Draw Lines, If At All?

Embryo research and manipulation take place within a constitutional framework protecting scientific inquiry and reproductive privacy.[57] The activities are presumed to be constitutional, in other words, and the government bears a heavy burden to justify limits on inquiry and reproductive choice. Governmental limits on embryo research are sporadic and ineffective. Nineteen states have fetal research laws that could be interpreted as restricting embryo research,[58] but the constitutionality of these laws has not been tested. The federal government's approach is to refuse to fund projects involving human embryos. This is a significant but not insurmountable limit in that researchers can look elsewhere for financial support. A law passed in one Australian state was so poorly drawn that it failed to provide useful guidelines for limiting problematic areas of research.[59] A bill to forbid embryo research was defeated in Britain.[60] Similar bills have been entertained in France and Germany,[61] but scientists in restrictive countries can arrange to visit facilities in other countries where embryo research is not substantively limited. In summary, embryo manipulations have no clear or systematic legal limits.

Two alternative approaches help meet public concerns about embryo manipulations. One is the use of interdisciplinary and cross-national commissions in which consensus is sought about the ethical acceptability of certain procedures.[62] Another is the development of voluntary guidelines by professional interest groups.[63] Neither approach has legal bite; each relies on professional mores and volunteerism. They may be useful, however, in encouraging voluntary restraints where legal limits are of questionable constitutionality. They also provide diverse forums for public discussion if they invite participants from different disciplines. The British have been the most creative in combining private guidelines with public authority to oversee embryo manipulations. The Statutory Licensing Authority is a cross-disciplinary group that reviews proposals for embryo manipulations and licenses centers that conduct the research.[64]

One question, then, is who should draw lines deemed appropriate—private organizations, public policymakers, or a combination of the two? The other is where the lines should be drawn, irrespective of who draws them. Ultimately, lines work best when there is a wide consensus that it is necessary to draw them. Lines already drawn seem to be temporary and subject to change, however. Several commissions, for example, have concluded that embryo research should not go beyond the fourteenth day of life,[65] but recent signs indicate a push to undo this line. American Fertility Society guidelines suggest that, "at this time" it is "prudent" not to conduct research on embryos beyond the fourteenth day. The Committee that developed the guidelines, however, notes that a strong case can be made for research beyond this point, and that further discussion is necessary on the matter.[66]

Arguably, the 14-day rule was easy to agree on because it represented no real sacrifice. It was discussed at a time when the ability to keep embryos alive for 14 days was a remote possibility. Moreover, researchers were unsure what to do with the embryos in any event. Now, at least one research group has suggested keeping embryos alive for longer than five or six days to study the properties of blastocysts.[67] Will the 14-day rule be abandoned when it appears embryos can be kept cleaving that long? How easily should consensual limits be modified when they deal with possible, not merely potential, acts?

Another ambiguous stopping point lies with embryo research itself. Early studies used a small number of spare or abnormal embryos from IVF. In one study, however, 181 normal and abnormal spare embryos were used in a controlled study.[68] The American Fertility Society Ethics Committee points out that the use of spare embryos may "adversely affect the reliability of the data obtained" and that the "production of human preembryos" may need to be undertaken to reach valid and reliable research conclusions.[69] Should lines be drawn on where embryos come from (spare, created with donor eggs and sperm), their status (normal or abnormal), their number, and the purpose of the research? Large-scale studies using created embryos offend societal sensibilities about the moral worth of embryos. On the other hand, the lack of controlled studies in IVF offend societal sensibilities about protecting couples from being continued experimental subjects in IVF. Should limits be placed on studies designed to promote the safety of embryo manipulations?

Conclusion

Embryo research is proceeding quietly in this country and more openly in Britain and elsewhere. Public discussions tend to revolve around the nature of the embryo. Although this is extremely important, it is only part of the issue before us. Published studies, advances in animal embryology, and a growing voice calling for embryo research tell us we are on the eve of clinical applications. We are not at present prepared for this development.

A first step in meeting the new issues is to publicize technical developments in embryo manipulations. What is possible?

What is now being offered at IVF clinics? What research studies are being planned? Open discussions about the goals, prospects, and desirability of embryo manipulations depend on realistic appraisals of what is now being done in laboratories and clinics. Sometimes, this means reading between the lines of technical reports. One ethics report referred to the possibility of teratogenesis tests on produced embryos, for example, and a scientific report ended with a mention of the possibility of exposing embryos to toxic membrane relaxants in pursuit of improved micromanipulation techniques. The people who draft these studies have research objectives in mind. What are they? Have they been widely discussed? How many people outside the IVF community are aware of these specialized visions for research?

A second step is to decide where embryo manipulation fits into current medicine. Is this a qualitatively unique field that warrants discussion as a separate entity, as argued in this chapter, or are embryo manipulations extensions of IVF and genetic inquiry and not appropriately discussed as a separate field? The answer to this question is critical in placing responsibility for embryo technologies. If embryo manipulations are simply extensions of IVF, then professional interest groups in that field, such as the American Fertility Society and the American College of Obstetrics and Gynecology, bear primary responsibility for developing guidelines with bite in the absence of governmental regulation. If embryo manipulations are conducted by geneticists and molecular biologists as extensions of genetic inquiry, on the other hand, responsibility for guidelines ought to rest with such organizations as the American Society of Human Genetics. If the embryo is rightly regarded as a "patient," then overarching medical organizations ought to take a leadership role in integrating standards of care into existing frameworks and modifying them where necessary.

Deciding where embryo manipulation fits into current medicine is necessary for filtering research reports to the public and for opening access for consumer, patient, women's, and public interest groups wishing to follow and question developments in the field. These groups also bear a responsibility for considering the unique dilemmas and issues raised by embryo manipulation. It does little good for embryo manipulations to become embroiled in the abortion debate, for example, when the issues are important enough to be discussed as discrete matters. An inopportune raising of the abortion issue has already dealt a possibly fatal blow to Congress' effort to create a Biomedical Ethics Board as a forum for discussing ethical matters relating to IVF and other topics.[70] Thus, just as practitioners have the responsibility for openly discussing their goals and discoveries, so do interest groups have the responsibility for publicizing and weighing as many sides of the issues raised by embryo manipulation as possible.

Gathering information about the state of embryo research is difficult in this country given the reluctance of researchers to speak publicly about what they perceive to be politically controversial activities. This secrecy is unfortunate. Conception is no longer "shrouded in mystery," and it is a throwback for clinicians to erect a new and more insidious shroud that precludes open debate and the crafting of principles to guide them as they offer embryo manipulations to infertile and genetically at-risk couples. New techniques and new dilemmas of reproductive choice are now upon us. Silence in matters with such profound implications for reproduction in the twentieth century raises perhaps the most deeply disturbing ethical problem of all.

Notes and References

[1]Joyce Carol Oates (1985) A terrible beauty is born. How? *New York Times Book Review*, August 11, p. 1.

[2]*See*, e.g., Jacques Testart, Bruno Lassalle, Robert Forman, Armelle Gazengel, Joelle Belaisch-Allart, Andre Hazout, Jean-Daniel Rainhorn, and Rene Frydman (1987) Factors influencing the success rate of human embryo freezing in an in vitro fertilization and embryo transfer program. *Fertility and Sterility* **48**, 107–112.

[3]Peter Aldhous (1990) Pro-life actions backfire. *Nature* **345,** 7.

[4]*See* Institute of Medicine (1989) *Medically Assisted Conception: An Agenda for Research,* National Academy Press, Washington, DC; Jan Tesarik (1989) Viability assessment of preimplantation concept: A challenge for human embryo research. *Fertility and Sterility* **52,** 364–366; CIBA Foundation (1986) *Human Embryo Research: Yes or No?* Tavistock Publications, London, UK; Embryo Research (1985), *Lancet* **i,** 2.

[5]*See* Andrea L. Bonnicksen (1989) *In Vitro Fertilization: Building Policy From Laboratories to Legislatures,* Columbia University Press, NY, pp. 75–82. The Department of Health, Education and Welfare disbanded the Ethics Advisory Board in 1979, which in effect imposed a moratorium on federal funding of research involving human embryos. According to federal law, such proposals must be reviewed by an ethics board before being considered for funding. Recent efforts to create a new EAB have been unsuccessful. *See* Andrea L. Bonnicksen (1989) Developments in the Status of Human Embryo Research: Revising the Policy Agenda, paper presented at the 1989 Annual Meeting of the Midwest Political Science Association, Chicago, IL.

[6]Yury Verlinsky, Eugene Pergament, and Charles Strom (1990) The preimplantation genetic diagnosis of genetic diseases. *Journal of In Vitro Fertilization and Embryo Transfer* **7,** 1–5.

[7]Howard W. Jones, Jr. and Charlotte Schrader (1989) And just what is a pre-embryo? *Fertility and Sterility* **52,** 189–191; Rafael I. Tejada and William G. Karow (1986) Semantics used in the nomenclature of in vitro fertilization, or let's all be more proper, *Journal of In Vitro Fertilization and Embryo Transfer* **3,** 341–342.

[8]Clifford Grobstein (1985) The early development of human embryos, *Journal of Medicine and Philosophy* **10,** 213–236, at 216.

[9]G. John Garrisi, Beth E. Talansky, Lawrence Grunfeld, Valdi Sapira, Daniel Navot, and Jon W. Gordon. Clinical evaluation of three approaches to micromanipulation-assisted fertilization, *Fertility and Sterility* **54,** 671–677, at 676.

[10]Jones and Schrader, p. 190.

[11]Ethics Committee of The American Fertility Society (1990) Ethical considerations of the new reproductive technologies, *Fertility and Sterility,* Supplement 2, **53,** 63S.

[12]Jones and Schrader, p. 189.

[13]Ethics Advisory Board, Department of Health, Education and Welfare (1979) *Report and Conclusions: HEW Support of Research*

Involving Human In Vitro Fertilization and Embryo Transfer, US Government Printing Office, Washington, DC.

[14]*See Davis* v. *Davis,* No. E–14496 (Blount County, TN, Equity Division [I], September 21, 1989 [equating the embryo with a child]); Instruction on respect for human life in its origin and on the dignity of procreation: Replies to certain questions of the day, Doctrinal Statement of the Vatican, March 10, 1987 (the human being is a "person from the moment of conception"); (1984) Ethical statement on in vitro fertilization, *Fertility and Sterility* **41,** 13 ("gametes and concepti are the property of the donors").

[15]This can take one of several forms: (1) remove one or more cells from a four- to eight-cell embryo to study, (2) divide the embryo in half to create monozygotic twins and study one-half of the embryo, or (3) excise the trophectoderm tissue (outer cell layer) of embryos at the later blastocyst stage. Verlinsky, Pergament, and Strom, pp. 2–3. So far, research has focused on the first method of examining one or two cells.

[16]J. L. Watt, A. A. Templeton, I. Messinis, L. Bell, P. Cunningham, and R. O. Duncan (1987) Trisomy 1 in an eight cell human pre-embryo. *Journal of Medical Genetics* **24,** 60–64; Roslyn R. Angell, A. A. Templeton, and R. J. Aitken (1986) Chromosomal studies in human in vitro fertilization, *Human Genetics* **72,** 333–339; Michelle Plachot, J. de Grouchy, Anne-Marie Junca, Jacqueline Mandelbaum, Catherine Turleau, P. Coullin, J. Cohen, and J. Salat-Baroux (1987) From ooctye to embryo: A model, deduced from in vitro fertilization, for natural selection against chromosome abnormalities. *Annales de Genetique* (Paris), **30,** 22–32; R. R. Angell, A. A. Templeton, and I. E. Messinis (1986) Consequences of polyspermy in man, *Cytogenetics and Cell Genetics* **42,** 1–7.

[17]Eve K. Nichols (1988) *Human Gene Therapy,* Harvard University Press, Cambridge, MA, p. 6.

[18]Dorothy Warburton (1987) Reproductive loss: How much is preventable? *New England Journal of Medicine* **316,** 158–160.

[19]Researchers have established a correlation between an embryo's appearance (whether it is fragmented or has one or three pronuclei instead of the normal two) and the incidence of chromosomal errors, showing a "high incidence of chromosomal errors in morphologically abnormal early preimplantation embryos." Sai Ma, Dagmar K. Kalousek, Christo Zouves, Basil Ho Yuen, Victor Gomel, and Young S. Moon (1990) The chromosomal complements of cleaved human embryos resulting from in vitro fertilization. *Journal of In Vitro Fertilization and Embryo Transfer* **7,** 16–21, at 16.

[20]Mark W. J. Ferguson, Contemporary and future possibilities for human embryonic manipulation, in *Experiments on Embryos* (Anthony Dyson and John Harris, eds.), Routledge, London, UK, pp. 6–26, at 7.

[21]Kate Hardy, Karen L. Martin, Henry J. Leese, Robert M. L. Winston, and Alan H. Handyside (1990) Human implantation development in vitro is not adversely affected by biopsy at the 8-cell stage. *Human Reproduction* **5,** 708–714, at 708. Researchers have also studied the RNA synthesis activities in human embryos. J. Tesarik, V. Kopecny, M. Plachot, et al. (1986) Activation of nucleolar and extranucleolar RNA synthesis and changes in the ribosomal content of human embryos developing in vitro. *Journal of Reproduction and Fertility* **78,** 463–470.

[22]Soon-Chy Ng, Ariff Bongso, Henry Sathananthan, and Shan S. Ratnam (1990) Micromanipulation: Its relevance to human in vitro fertilization. *Fertility and Sterility* **53,** 203–219, at 214; Ferguson, p. 7.

[23]C. Coutelle, C. Williams, A. Handyside, K. Hardy, R. Winston, and R. Williamson (1989) Genetic analysis of DNA from single human ooctyes—A model for preimplantation diagnosis of cystic fibrosis, *British Medical Journal* **299,** 22–24; Verlinsky, Pergament, and Strom, p. 3.

[24]Hardy, Martin, and Leese. The agency was known as the Interim Licensing Authority when this project was approved.

[25]They concluded that the biopsy, while reducing the cellular mass, does not adversely affect the preimplantation development of biopsied embryos in vitro. *Ibid.*

[26]*Ibid.,* p. 711. *See also* Virginia N. Bolton, Susan M. Hawes, Clare T. Taylor, and John H. Parsons, Development of spare human preimplantation embryos in vitro: An analysis of the correlations among gross morphology, cleavage rates, and development to the blastocyst, *Journal of In Vitro Fertilization and Embryo Transfer* **6,** 30–35, at 35.

[27]Ethics Committee of the American Fertility Society, p. 63S; Ferguson. Twinning will raise multiple questions. Should it be used regularly as a back-up in IVF? As a way of giving a couple the pleasure of twins but without having them at the same time?

[28]Zona drilling is done by applying acid, by mechanical means with a microneedle, or by variations of either of these techniques. Garrisi et al., 67–72.

[29]S. C. Ng, T. A. Bongso, S. S. Ratnam, Henry Sathananthan, Clement L. K. Chan, P. C. Wong, Leiff Hagglund, C. Anandakumar, Y. C. Wong, and Victor H. H. Goh (note)(1988) Pregnancy after transfer of multiple sperm under the zona. *Lancet* **ii,** 790.

[30]Ismail Kola, Orly Lacham, Robert P. S. Jansen, M. Turner, and A. Trounson (1990) Chromosomal analysis of human oocytes fertilized by microinjection of spermatozoa into the perivitelline space. *Human Reproduction*, **5**, 282–285.

[31]Jon W. Gordon, Larry Grunfeld, G. John Garrisi, Daniel Navot, and Neri Laufer (1989) Successful microsurgical removal of a pronucleus from tripronuclear human zygotes. *Fertility and Sterility* **52**, 367–372.

[32]Henry E. Malter and Jacques Cohen (1989) Embryonic development after microsurgical repair of polyspermic human zygotes. *Fertility and Sterility* **52**, 373–380, at 373.

[33]*Ibid.*

[34]Gordon. For unsuccessful attempts, *see* references listed by Gordon.

[35]*Ibid.*

[36]Jacques Cohen,Carlene Elsner, Hilton Kort, Henry Malter, Joe Massey, Mary Pat Mayer, and Klaus Wiemer (1990) Impairment of the hatching process following IVF in the human and improvement of implantation by assisting hatching using micromanipulation. *Human Reproduction* **5**, 7–13, at 7.

[37]Barbara F. Culliton (1990) Gene therapy begins. *Science* **249**, 1372.

[38]Ferguson, p. 11.

[39]Harold M. Schmeck, Jr. (1982) Rat gene implant in mice reported. *New York Times*, December 16, p. 1. *See also* Jeremy Cherfas (1990) Molecular biology lies down with the lamb. *Science* **249**, 124–126.

[40]Janice Sharp (1991) The patenting of transgenic animals, in *Emerging Issues in Biomedical Policy: Annual Volumes* (Robert H. Blank and Andrea L. Bonnicksen, eds.), Columbia University Press, NY, (in press).

[41]Robert B. Church, ed. (1990) *Transgenic Models in Medicine and Agriculture: Proceedings of a UCLA Symposium Held at Taos, New Mexico, Jan. 28–Feb. 3, 1989*, Wiley-Liss, Inc., NY.

[42]W. French Anderson (1990) Genetics and human malleability. *Hastings Center Report* **20**, 21–24.

[43]*Ibid.*

[44]National Institutes of Health (1986) Points to Consider in the Design and Submission of Human Somatic-Cell Gene Therapy Protocols, September 29, reprinted in Nichols, pp. 195–208.

[45]*See,* generally, Ethics Advisory Board; CIBA Foundation; Anne McLaren (1987) Can we diagnose genetic disease in pre-embryos?

New Scientist, December 19, pp. 42–45; John A. Robertson (1986) Embryo research. *University of Western Ontario Law Review* **24,** 15–37; Hans-Martin Sass (1985) Moral dilemmas in perinatal medicine and the quest for large scale embryo research: A discussion of recent guidelines in the federal republic, *Journal of Medicine and Philosophy* **10,** 279–290.

[46]E. Joshua Rosenkranz (1987) Custom kids and the moral duty to genetically engineer our children. *High Technology Law Journal* **2,** 1–53, who argues there may be a moral obligation to rescue an ailing embryo if it is possible to do so.

[47]For example, the American Fertility Society concludes that research embryos must be treated with respect and embryos to be transferred to the woman's uterus be treated with concern beyond respect. Ethics Committee, pp. 32S–33S.

[48]*See,* generally, Instruction; Paul Ramsey (1972) Shall we 'reproduce'? *Journal of the American Medical Association* **220,** 1346–1350, and 1480–1485; Christine Ewing (1989) The case against embryo experimentation: A feminist perspective. *Legal Service Bulletin* **14,** 109–112, 121.

[49]David Jabbari (1990) The role of law in reproductive medicine: A new approach. *Journal of Medical Ethics* **16,** 35–40.

[50]Warburton.

[51]Verlinsky, Pergament, and Strom, p. 4.

[52]*See ibid.,* p. 4, who wonder if biopsies should become the standard of care for all couples for whom advanced maternal age is a factor.

[53]Office of Technology Assessment (1988) *Infertility: Medical and Social Choices,* OTA-BA-358, US Government Printing Office, Washington, DC, 150–151.

[54]Andrea L. Bonnicksen (1988) Embryo freezing: Ethical issues in the clinical setting. *Hastings Center Report* **18,** 26–30.

[55]Ferguson, pp. 8–9.

[56]Kate Hardy, Alan H. Handyside, and Robert M. L. Winston (1989) The human blastocyst: Cell number, death and allocation during late preimplantation development *in vitro. Development* **107,** 597–604, at 598.

[57]*See* Ira H. Carmen (1981) The constitution in the laboratory: Recombinant DNA research as "Free Expression." *Journal of Politics* **43,** 737–762; *Roe* v. *Wade,* 410 US 113 (1973).

[58]Office of Technology Assessment, p. 251. For example, Louisiana passed a law defining the embryo as a juridical person with the power to sue and be sued. La. Rev. Stat. 14:87,121–133.

[59]Margaret Brumby and Pascal Kasimba (1987) When is cloning lawful? *Journal of In Vitro Fertilization and Embryo Transfer* 4, 198–204.

[60]CIBA Foundation, p. 3; (1987) Draft legislation on infertility services and embryo research. Lancet, **ii,** 1343.

[61]David Dickson (1989) France introduces bioethics law. *Science* **243,** 1284; Richard Sietmann (1990) Abortion divides uniting Germanies. *Science* **249,** 1100.

[62]For a summary of commission reports, *see* LeRoy Walters (1987) Ethics and new reproductive technologies: An international review of committee statements. *Hastings Center Report* **17,** 3S–9S.

[63]Bonnicksen, *In Vitro Fertilization,* pp. 82–90.

[64]John Warden (1990) Lords approve embryo research. *British Medical Journal* **300,** 416.

[65]*See* Ethics Committee, p. 62S, for citations.

[66]*Ibid.,* p. 63S.

[67]Hardy, Handyside, and Winston. The group kept 181 normal and abnormal spare embryos from IVF cleaving for up to seven days (to up to 125 cells) in order to understand more about the process in which the human embryo expands into two types of cells at the blastocyst stage. The study was conducted to learn more about the early development of the human embryo in culture, p. 601. The researchers ended their published report with a statement of the need for further study.

[68]*Ibid.*

[69]Ethics Committee, p. 63S.

[70]Bonnicksen, Developments.

Prenatal Diagnosis

Barbara Katz Rothman

The typical medical ethics case, as it is standardly presented for discussion, involves Doctor Goodguy sitting at his desk, when in walks Patient Problem, presenting an ethical dilemma. In the area of prenatal diagnosis, the doctor is typically an obstetrician or a geneticist, the patient is a pregnant woman, and the dilemma is abortion. Very often, the case involves a woman who wants an abortion for a reason the doctor feels to be insufficient (with all the assumptions about sufficiency, abortion, fetuses, and women included). Occasionally, it involves a woman who refuses testing the doctor thinks necessary (again, with assumptions about necessity and all the rest included).

Example 1: Mrs. X (we are told) comes from a culture that strongly values sons. She is the mother of four daughters, and pregnant again at the age of 37. She requests amniocentesis, and since current medical thinking regards a woman over age 35 as "high risk" for Down Syndrome, she is considered to be "entitled" to this amnio. However, she makes it clear to the physician that a healthy girl would be just as unacceptable as would a child of either sex with Down Syndrome. Should Doctor Goodguy do the test? Can the doctor refuse to divulge information about fetal sex?

From: *Biomedical Ethics Reviews • 1991*
Eds.: J. Humber & R. Almeder ©1991 The Humana Press Inc., Totowa, NJ

Example 2: Mrs. Y learns through a "routine" sonogram that her fetus has a limb deformity. She reacts with repulsion and horror, and wants the prenancy aborted at once. Should Doctor Goodguy do the abortion?

Example 3: Miss Z (and this one is indeed most often presented as "miss") presents a different dilemma. She is not interested in the testing or the abortion being offered. Miss Z is a 22-year-old former or current user of IV drugs, and so, at risk for HIV infection. She either refuses HIV testing, or if tested and found positive, she decides to continue the pregnancy and risk the roughly one-in-three chance that her infant will develop AIDS.

First, it is important to note that each of these scenarios assumes that the dilemma arrives with the patient. That is probably because biomedical ethics as a field is often closely associated with hospitals, medical schools, and physicians. From the perspective of the patient, of course, the dilemma arrives with the doctor. A woman living in a culture that values sons over daughters is suddenly offered a test for fetal sex, creating an element of choice and responsibility, and thus, a dilemma. Or, in the second or third example, with much effort to "penetrate the maternal barrier" and "access the fetal patient," physicians developed, and then began to use routinely, ultrasonography. So, a woman arriving for routine prenatal care suddenly learns distressing things about her fetus—things she may not have asked to learn. Tests are developed for HIV status and then offered to women. With the offering of the test comes the dilemma for the woman.

There is, however, a more profound critique to make of these three examples and the assumptions that they embody. Each of these examples encourages us to focus on the level of individual decision making. We ask ourselves—and our students—what should Doctor Goodguy do? What should Patients X, Y, or Z decide? Often, the problem is presented to students with the five principles of autonomy, veracity, beneficence, nonmaleficence, and justice offered as guides to finding the best answer.

As a social scientist, I have a strong need to shift the focus. Dilemmas of this sort do not simply "arise" as if they were spontaneously generated, nor do they reside in either of the individuals, patient or doctor, presented as in conflict. These dilemmas are socially, politically, and economically constructed. This is particularly clear with the dilemmas presented by new technology. The institutionalization of new technologies does not occur on the individual level, and is not the work of individual inventors and consumers. We must move considerably past questions of "choice" to understand the dilemmas that confront us.

The new reproductive technologies, including and perhaps especially, the techhologies of prenatal diagnosis—amniocentesis, sonography, chorionic vullus sampling, preimplantation diagnosis, and more to come—are offered to people in terms of expanding choices. However, it is always true that although new technology opens up some choices, it closes down others. The new choice is often greeted with such fanfare that the closing of the door on the old choice goes unheeded. To take a simple example, is there any meaningful way one could now choose horses over cars as a means of transportation? The new choice of a "horseless carriage" eventually left us "no choice" but to live with the pollution and dangers (as well as the convenience and speed, of course) of a car-based transportation system.

In the area of reproductive technologies, this closing down of choice happened first with the quantity of children. The oldest and most basic reproductive technology is the technology of fertility limitation. Self-imposed limits on fertility, through contraception, abortion, or a combination of the two, are the sine qua non of the reproductive rights movement, and yet, we must realize that the choice of contraception simultaneously closed down some of the choice for larger families. North American society is geared to small families, if indeed, to any children at all. Without the provision of good medical care, day care, decent housing, and schooling, children are luxury items; fine if you can afford them.

And so, it may be also with prenatal diagnosis, which serves as a technology of quality control, based on a given society's ideas about what constitutes "quality" in children. The ability to control the "quality" of our children may ultimately cost the choice of not controlling that quality. Individual families, but most especially individual mothers, bear the costs of children and the special costs of special children. How much any given child costs a mother is based not only on the condition of the child, but even more on the conditions of the mother's life. Any analysis of a woman's choice of any abortion, but most especially a selective abortion, has to recognize the context in which the decision to abort is made, and the circumstances in which the woman is placed. As Rosalind Petchesky has stated:

> The "right to choose" means very little when women are powerless. ...Women make their own reproductive choices, but they do not make them just as they please; they do not make them under conditions which they themselves create, but under social conditions and constraints which they, as mere individuals, are powerless to change.[1]

We live in a system in which women and children are both disvalued, an antichild-antiwoman society. It is women and children who are poor, whose needs are not being met. In this system, women and children are often pitted against each other, competing for scarce resources. The mother finds herself becoming a resource: Her own life (and specifically, her own time) is to be divided between herself and her children. Whatever the children get, it may very well be coming off the life of the mother—in time, in attention, in emotional support, sometimes in food and basic necessities. It is in this context that mothers are judged in terms of their willingness to sacrifice. The more she gives of herself to her children, the better a mother the society says she is. The more she holds back of herself, for herself, the more she runs the risk of being the "wicked stepmother," evil in her selfishness.

When women and children are both disvalued, to speak for the rights of either, to the needs of either to be met, is then to contribute to the disvaluation of the other. When one adds to the situation the virtually total disvaluation of the needs of the disabled, the "defective" or "invalid" people, the place of selective abortion in our society is highlighted. Women know that children with "special needs" make special demands. The society as a whole has shown itself unwilling to meet those demands—we are, as a society, unwilling to meet the ordinary needs of ordinary children. With wonderful and notable exceptions, fathers and other family members have not risen to the occasion. The burden of childrearing, of all childrearing, has fallen overwhelmingly on individual mothers. Although those in the disability rights movement rightly resent the use of the word "burden" to describe their lives, it is not a description unique to the disabled. Children, all children, can be described as burdensome when their needs fall almost exclusively on one person. Yes, they are also delightful, joyous, pleasures, and treasures, whether able-bodied or disabled, but side-by-side with the pleasures come the sacrifices. The individual women, or at the very best, the individual couple or family, can demand more and more from the society for the child, and, in fact, making such demands becomes one of the chief responsibilities of the parents of a disabled child. It is clear to us all, however, that the society will not respond with openness and generosity, and most assuredly cannot be depended on to continue responding to the child's needs when the mother is no longer there. Even if the woman were to be willing to sacrifice herself entirely to meet the needs of the child, it may still not be enough.

It is, in this context, that amniocentesis and the other technologies of prenatal diagnosis, permitting selective abortion, are introduced, giving an illusion of choice, allowing individuals to believe that they have gained control over the products of conception. However, the choices are made within an ever-narrowing structure. Issues of basic values, beliefs, and the larger moral questions will be lost in this narrowing of choices as decisions become pragmatic, often clinical, and always individual. Irving Kenneth Zola puts it this way:

> Bombarded on all sides by realistic concerns (the escalation of costs) and objective evidence (genetics) and techniques (genetic counseling), the basic value issues at stake will be obfuscated. The freedom to choose will be illusory. Someone will already have set the limits of choice (cuts in medical care and social benefits but not in defense spending), the dimensions of choice (if you do this then you will have an x probability of a defective child) and the outcomes of choice (you will have to endure the following social, political, legal and economic costs).[2]

Thus, the new technologies of prenatal diagnosis and selective abortion do, indeed, offer new choices, but they also create new structures and new limitations on choice. Because of the society in which we live, the choices are inevitably couched in terms of production and commodification, becoming matters of "quality control." The dilemmas then get seen in terms of choices about quality of life for the individuals and potential individuals involved.

It is with this context in mind that I turn now to the experiences of women with prenatal diagnosis and selective abortion. Because, whereas the social, economic, and political context is too often missing from discussions of biomedical ethics, so too is the voice of the lived experience.

Therefore, it is important to hear the actual words of women facing the decision to terminate a pregnancy following prenatal diagnosis. The women quoted below were interviewed as part of a larger study on women's experiences with prenatal diagnosis and selective abortion.[3] I will focus here on the issue of responsibility as experienced by women who had prenatal diagnosis and learned that there was a serious problem with the fetus. These women are in the nexus between the society that largely creates or structures the problem—the profound cutting of already limited services for disabled children and adults that we have seen over the decade, the deep stigma attached to disability; the privatized and relatively isolated nuclear family; the gender inequality that leaves women uniquely bearing the costs of their children's disabilities—and the technology that is supposed to solve the problem.

The Tentative Pregnancy

How does anyone decide whether to continue or to terminate the pregnancy when given a bad diagnosis? The overwhelming majority of women who get a bad diagnosis do terminate. In part, that is because most of the women who would choose not to terminate sensibly avoid having the tests and facing their decisions. Even though the decision to have the amniocentesis implies, for most women, the willingness to abort for a bad diagnosis, the actuality of the diagnosis often requires that the decision be made anew. There are several reasons for this. Some women are pressured into the amniocentesis. They sometimes give in because the chances of a bad diagnosis are so remote that it is easier to go along with it than to argue with husband or doctor. Some women postpone the decision purposely, wanting to "get the information, and then decide." Some actively seek out amniocentesis, fully expecting to terminate for a bad diagnosis, but find themselves more deeply affected by the pregnancy itself than they had expected. The lateness of the results changes their readiness to abort, and they have to decide again. And some women get unexpected diagnoses, not the bad news for which they were prepared, but other, surprising bad news, requiring new decisions.

It is generally agreed that the most straightforward decision making occurs when the fetus is diagnosed as having a fatal condition. If the fetus is going to die at birth, then there is often understood to be "no point" in continuing the pregnancy. Six of the thirteen women I discuss in detail in this chapter were in, essentially, that position. Laura carried an anencephalic fetus, a fetus without sufficient brain development for it to survive more than days after birth. Laura herself, however, remains unsure that the condition was inevitably fatal. The rest of these six women were convinced that their fetus had no chance. Fern's fetus had a diagnosis of a genetic kidney disease and a 99% chance of dying even before the end of the pregnancy. Donna's fetus had a blood and bone disease in which the baby bleeds to death. Her third child had the disease and died at five weeks. This was her fourth pregnancy. Shirley never got a clear diagnosis, but it was obvious that her fetus was dying and making her very sick as well. She aborted to save her own life. Denise carried a fetus with Trisomy 18 and spina bifida. She was told it had a small chance of survival and would lead only a minimal kind of existence at best. Andrea's fetus had Tay Sachs disease, which kills not *in utero* or in infancy, but does invariably kill within the first years of life. The remaining seven women carried fetuses with the extra twenty-first chromosome that is Down Syndrome. Is it any easier for those women whose fetuses would die anyway? I am not sure. That knowledge does not take away the sense of responsibility the women feel. I specifically say responsibility, and not guilt. Some women express feelings of guilt, but all of them express "the inescapable sense of deep responsibility." Listen to Fern, who knew her fetus was dying of kidney disease:

> There are times that I really curse modern technology. No one should have to make these kinds of decisions. There are occasional flukes of nature where things don't work out, or at least they don't seem to, and yet I very firmly believe that there is always something good to be found in every situation, no matter how grim. I also think that most women know in their hearts whether or not their baby is going to be normal, and that emotionally they are prepared for it before the baby's born.

Having articulately expressed the same feelings and beliefs held by so many of the women who refused amniocentesis, Fern, at this point, apologizes for "rambling" and being "not coherent." She goes on to say that it is an individual matter, and each time the decision needs to be made, all circumstances need to be evaluated. It is not that Fern feels she should not have done what she did. She is angry at the local hospital that refused her admission for what they called "abortion on demand," making her seek an outpatient facility:

> They were bound and determined for us to have this baby regardless of the pain and suffering he would have to endure prior to and after his birth.

Making the baby suffer would be wrong. The abortion had to be done, yet she wished that "the baby would just hurry up and die so that we wouldn't have to murder it first." She compared the abortion with a spontaneous miscarriage she had, saying it is the same because "it is a life lost and with it all the hopes and dreams of that new being." However, the spontaneous miscarriage was different, "because I was not the one that actively murdered my baby." With the abortion, "I had good reason for doing it, but it was still a conscious decision to end that life."

Andrea, who terminated for Tay Sachs disease, also expresses this sense of responsibility; not guilt, but certainly responsibility:

> This is your responsibility. You have to make the choice. No one makes that choice for you.

However, even believing that the fetus would die does not protect entirely from guilt. While all the women echo the theme of choice and responsibility, Denise remains the most troubled. Though she thought the amniocentesis was "a very intelligent thing," she says:

> In retrospect, I have wondered if it might have been easier
> on me just to carry the pregnancy to term and lose the child
> that way. I think emotionally it might have been—well, that's
> guessing. Maybe it would just have dragged it out for a
> longer time and made it just as hard if not harder, for a longer
> period. Yeah, this was a real person to me, and all the ra-
> tionalizing in the world is not going to change my feeling.
> But my husband doesn't consider that the baby was a person.

Denise comes back again and again to the ultimate responsibility
in the decision: "An abortion is a choice that you make and despite what other people say to you, it's ultimately your own choice,
it's something you do, and—I kind of feel like I committed a
murder."

When the fetus will suffer and die, then the abortion can be
seen as a painful obligation the mother has toward her fetus,
toward her baby. Even Denise wonders only whether it might
have been easier on herself had she continued the pregnancy "and
lost the child that way." Fern, fighting her local hospital for admission for the abortion, is no different than the mothers of infants in neonatal intensive care units who fight to let their suffering
babies die in peace. A sacrifice is occurring, suffering is happen-
ing, but in many ways it is the socially accepted sacrifice in
mothering: The mother suffers to spare her child.

The question of responsibility and obligation, of choice and
of sacrifice, becomes more complicated when the diagnosis is
Down Syndrome. As Eleanor describes the dilemma:

> The baby can live to a mature age, and have a rather good
> life, so there's a tremendous amount of guilt involved—that
> you're getting rid of it because it is not a perfect human
> being—and it's your decision—it's not God's decision, or
> nature's decision, it's yours and yours alone, so it carries
> with it a heavy weight.

The responsibility, when the diagnosis is a fatal condition,
is the responsibility for determining the timing and the mode of

the baby's death. With Down Syndrome, the responsibility is more directly one of life and death.

No one can predict in detail the condition of the baby just by seeing the extra twenty-first chromosome. Some women are left wondering if the baby might have been only mildly retarded. One of the sadder ironies in the diagnostic process, however, is that some of the fetuses that are aborted would have miscarried shortly, and some would have died in infancy. The woman has the experience of shouldering responsibility for a decision that "God or nature" might have made for her. Sondra had the comfort of learning that the fetus she carried did have many physical problems:

> It made me feel better in a sense, that it wasn't just Down Syndrome, but heart and lung damage. I never read the autopsy report; [the doctor] told us that, and it made me feel better.

It may have made her feel better, but:

> The first few months were really horrible. ...Guilt feelings, feelings of emptiness—it was terrible. My husband was really great—he had to drag me out of bed, I would just lie in bed. Didn't want to talk to anybody, didn't want to move, really didn't want to do anything.

Sondra did not know about the heart and lung problems before she aborted; she had to come to a decision based on only the information that the fetus had Down Syndrome. She thought about the kind of life the world would offer her baby:

> I have a handicapped sibling and I'm very conscious of how our society deals with the handicapped. I couldn't in right conscience at that time decide to bring a handicapped individual into the world. It's a tremendous decision.

Sondra is painfully aware of the seeming contradiction that exists between her commitment to the rights of the disabled, and the decision she made. "On the one hand you say the handicapped deserve all these rights and then on the other hand say that this child doesn't deserve to live." In addition to her experience with her sister, who had polio as a child, Sondra teaches emotionally disturbed children:

> I had trouble dealing with the kids, sort of like, "They're alive, and they're going to go out and kill somebody one day, half of them, and Down Syndrome people don't. They're not doers, they just need a lot, they don't really take a lot." I was angry.

Even more than Sondra, Beryl understood what the world held in store for her fetus with Down Syndrome, should she have continued the pregnancy. Beryl has a Master's degree in special education, specializing in mental retardation, and worked for thirteen years teaching the moderately, severely, and profoundly retarded:

> How ironic to choose to terminate a pregnancy which, left to nature, I was thoroughly prepared to cope with! ...I told my geneticist I almost envied the relatively uninformed who could conjure up an image of the fat retardate on the street corner, mouth sagging, etcetera, and make their decision in the recoil. I think being aware of the tremendous steps which have been made with the retarded was less than an asset to me at times and merely introduced more irony into an already complex decision.

So, how did Beryl come to the decision to terminate a pregnancy for Down Syndrome? She does not "recoil" from the retarded; she clearly takes great satisfaction in her work and plans to continue in it. It seems, for Beryl as for Sondra, it is not what she knew about the fetus that determined the decision, but what she knows about our world:

> If all of society—including extended family—shared the enthusiasm and confidence in the retarded that we in my

work field share, decisions such as ours would be fewer. ...When I read accusations of being like the Nazis, having no room for anyone but the "perfect," etcetera, I sizzle. ...Actually, if I were the only one involved, I would have kept the baby and used the best of my training to raise him. But to me the burden placed on the rest of the family, and on society, as I age or die, and the burden which in turn would fall upon the child, is too great to justify satisfying my ego.

Is this really any different than the decision that the other women faced, those whose fetuses were going to die? There is going to be pain and suffering, there is going to be a sacrifice made. Once again, the woman chooses to take on the burden herself, to bear the responsibility for the choice. She lives with the pain of her choice, with her grief and loss, to spare her child.

However, it is not only their babies that they spare. Whereas the abortion calls forth pain and grief, so too does the experience of mothering a child with Down Syndrome. Knowing that the sacrifice of the fetus has not only cost the mother grief, but also spared her other grief, becomes a source of guilt. Since "goodness" for a man is often measured by her willingness to give of herself,[4] can she be sure of her own motives in this complex decision? Sacrificing self for child is "good," sacrificing child for self, "wicked."

Guilt looms large for Anna. Months after her abortion, her car swerved as she and her husband drove in the rain. The thought flashed through her mind: "If we get killed, we deserve it." She sees what she avoided by aborting, and thinking about that, says she decided "for very selfish reasons...[to] take the choice encouragingly offered to me to abort the kid." Having made the choice, having terminated the pregnancy, "I felt sad but resigned, and a bitterness about these terrible choices set in that I'm still shrouded in. ...I feel brittle, with an icy sheath around my heart. I'm on guard."

Elisabeth too speaks of guilt, and openly questions the rightness of her decision. She had the amniocentesis because:

> It seemed unkind to knowingly bring a Downs child into the
> world, and unkind to not find out if the possibility existed.
> There was no education for the hell of waiting for the results,
> or the excruciating continuing sadness and guilt at killing
> our child.

Elisabeth did what she did out of kindness, but now:

> The question I wonder about now is, is our assumption cor-
> rect that it is unkind to knowingly bring a Downs child into
> the world. The Downs child can know and express love and
> joy and pain. Isn't this enough? By killing a Downs child I
> bow before the false god of intelligence. Isn't intelligence
> overvalued in our society at the expense of other values?

She remains isolated in her guilt and grief she has never been
able to talk with anyone who has made a similar decision. When
I thank her for giving of herself to me, she says:

> I would talk with anyone, anytime, who has questions. It was
> the saddest, most guilt-producing anguish of my life.

Unlike those who seek comfort in another baby, Elisabeth
and her husband "gave up our dream of a biological addition to
our family." Her husband had a vasectomy. She did not want to
face these choices again:

> I do not want to terminate the life of another child. I do not
> want to bring a Downs child into the world. I will not have
> another pregnancy.

It still seems wrong to Elisabeth to bring into the world a child the
world so clearly does not want and will not care for. When both
having the child and not having the child would be wrong, guilt
is inescapable.

What does it do to a person to make a monumental decision, and not ever be able to be sure of its rightness? Heather says:

> I don't know what it is, but you lose a little bit of your beliefs—it changes you in many ways. Your outlook on life changes, the moral concept—I would say it affects everything. I'm not going to say what I would have been two years ago.

Heather had the abortion because it was the "mature" thing to do. She was visiting relatives when the results came:

> You go through all stages, crying and resentment, it cannot be. ...Some nights were sleepless, you toss and turn, angry, and then something tells you just be mature and approach it from the mature point of view. And so I just packed up my suitcase and went home. Sunday I flew in and Monday morning I was already in the hospital.

Maturity, morality, religion—all of her basic values were challenged:

> Morally, there is a question—you have to battle with yourself. ...Here we talk about life, what is life, do we have a right—...[My religious beliefs] changed dramatically. In fact, I drew away from the church. Because how can I justify myself? ...You never forget, your life is never the same, but it's still a life. Sometimes it seems like a movie, it just happened, it's not affecting you. My life is straightened out now.

With all of her own anguish, Heather still supports amniocentesis.

> I would only encourage the amniocentesis even though I do not think it is the most accurate testing and there is a lot of pain involved too—and—the pain you cannot really describe. ...It's a grieving process. ...And emotional turmoil there's a lot of whys asked, and questions asked. But I guess if you feel secure in your relationship and your environment, you

can overcome it. You can cope with it, too. But you don't believe it's happened to you—it stays with you and you feel like you're a victim.

A victim. These women are not the villains some would have us believe, aborting fetuses it would be inconvenient to raise, searching for the "perfect" child. They are the victims. They are the victims of a social system that fails to take collective responsibility for the needs of its members, and leaves individual women to make impossible choices. We are spared collective responsibility because we individualize the problem. We make it the woman's own. She "chooses," and so we owe her nothing. Whatever the cost, she has chosen, and now it is her problem; not ours.

References

[1]Rosalyn Petchesky (1980) Reproductive freedom: Beyond a woman's right to choose. *Signs: Journal of Women in Culture and Society* **5,** 661–685.

[2]Irving Kenneth Zola (1983) *Socio-Medical Inquiries: Recollections, Reflections and Reconsiderations,* Temple University Press, Philadelphia, PA, p. 296.

[3]Barbara Katz Rothman (1986) *The Tentative Pregnancy: Prenatal Diagnosis and the Future of Motherhood,* Viking Press, New York, NY.

[4]Carol Gilligan (1982) *In a Different Voice: Psychological Theory and Women's Development,* Harvard University Press, Cambridge, MA.

Author Index

Subject Index